World Review of Nutrition and Dietetics

Vol. 104

Series Editor

Berthold Koletzko

Dr. von Hauner Children's Hospital, Ludwig-Maximilians University of Munich, Munich, Germany

Richard D. Semba

The Vitamin A Story

Lifting the Shadow of Death

41 figures, 2 in color and 9 tables, 2012

Basel · Freiburg · Paris · London · New York · New Delhi · Bangkok ·
Beijing · Tokyo · Kuala Lumpur · Singapore · Sydney

Dr. Richard D. Semba
The Johns Hopkins University
School of Medicine
Baltimore, Md., USA

Library of Congress Cataloging-in-Publication Data

Semba, Richard D.
 The vitamin A story : lifting the shadow of death / Richard D. Semba.
    p. ; cm. -- (World review of nutrition and dietetics, ISSN 0084-2230
; v. 104)
 Includes bibliographical references and index.
 ISBN 978-3-318-02188-2 (hard cover : alk. paper) -- ISBN 978-3-318-02189-9
(e-ISBN)
 I. Title. II. Series: World review of nutrition and dietetics ; v. 104.
0084-2230
 [DNLM: 1. Vitamin A Deficiency--history. 2. History, 19th Century. 3.
Night Blindness--history. 4. Vitamin A--therapeutic use. W1 WO898 v.104
2012 / WD 110]

 613.2'86--dc23

                                                    2012022410

Bibliographic Indices. This publication is listed in bibliographic services, including Current Contents® and PubMed/MEDLINE.

Contents

Dedication

For Rita

Preface

My early experience in international health coincided with the fitting of the last piece into the centuries-old vitamin A puzzle. Understanding of these vital food components was only beginning to come into focus when the word in its original form – vitamine – was coined by Polish-American biochemist Casimir Funk in 1912.

As a twenty-five-year-old medical student in 1980, I worked with a Venezuelan medical team treating victims of river blindness (onchocerciasis), a parasitic infection spread among humans by black flies. River blindness was then a leading cause of blindness worldwide, known to afflict nearly 20 million people. Our Venezuelan patients were Yanomami Indians living in the remote headwaters of the Upper Orinoco River. The team was charged with administering intravenous injections of suramin to river blindness victims. (Suramin was developed in pre-War Germany as Bayer 205. It is still in use to treat sleeping sickness.) Suramin can produce nasty adverse reactions, including fever, nausea, rash, and headaches. In extreme cases, a patient can collapse and die during suramin treatment. The team's nurses were so fearful of causing harm that some inserted the needle in a patient's vein but then withdrew the syringe without ever injecting the medication.

The need for a safer river blindness treatment lasted only a few more years. In 1984, while in my residency training in ophthalmology, I joined a scientific team in Liberia studying river blindness at the Uniroyal Rubber Plantation, where the disease afflicted many of the rubber tappers. My colleagues were conducting a clinical trial to see whether ivermectin, a versatile drug with both veterinary and human uses, was effective in treating river blindness. We were pleased to find ivermectin highly effective in both treating the disease and its complications without dangerous side effects. Ivermectin is now the mainstay river blindness treatment and is given communitywide in places wherever river blindness occurs – one tablet, once a year. As a result, this onetime scourge is now under control.

Having participated in the river blindness/ivermectin success, I wanted to tackle another, harder, ophthalmological problem. I was able to do this in 1987, after completing my training at the Wilmer Eye Institute at John Hopkins. With a Physician-Scientist Award from the National Institutes of Health, I decided to work on the particularly persistent problem of vitamin A deficiency, which was known to be a leading cause of blindness and death among developing countries' children and a major cause of illness and death in childbearing women. Alfred Sommer, then a professor of ophthalmology and later dean of the Johns Hopkins School of Public Health, encouraged me to join the efforts to understand and control vitamin A deficiency.

It was an exciting moment in public health, with signs of progress on the horizon, but also with frustrating questions still looming. Studies were beginning to suggest that oral doses of vitamin A, when given to young children, could protect them against diarrhea, measles, blindness, and death. Exactly how that worked, however, remained unknown.

I began my first work on vitamin A in Indonesia with Muhilal, a nutritionist (like many Indonesians, Muhilal uses one name only), and Gantira Natadisastra, an ophthalmologist and director of the Cicendo Eye Hospital in Bandung. Our research found that children living on vitamin A-poor diets had weakened immune systems, which went part way toward explaining why they were particularly susceptible to infectious diseases. Looked at the other way, vitamin A was emerging as essential for the proper function of the immune system.

Research groups elsewhere were finding corroborating evidence that vitamin A deficiency weakens immunity. A consensus was growing that vitamin A deficiency is, in fact, an acquired immune deficiency disorder. As such, it can be categorized along with AIDS, but only partly, because its cause is not viral but nutritional. On the one hand, vitamin A deficiency greatly increases susceptibility to infections, many of them potentially fatal such as measles, diarrhea, dysentery, and tuberculosis. One the other hand, once understood, vitamin A deficiency is tractable in ways that AIDS is not. Adequate intake of vitamin A can enable the body to resist – and overcome – these infections. In other words, treatment with vitamin A can cure the conditions that deficiency caused.

These recently determined attributes of vitamin A – that it can both prevent and cure – have placed it at the top of the international public health agenda. Vitamin A supplementation has become part of the basic public health canon of interventions to improve child survival. The other fundamental public health interventions are childhood immunizations against common killers such as tetanus, diphtheria, whooping cough, polio, and measles; iodized salt to prevent goiter and cretinism; oral rehydration to counter the potentially fatal effects of diarrhea; and clean water and sanitation to reduce the spread of dysentery, cholera, and other water-borne illnesses. Vitamin A supplementation alone has saved the lives of an estimated 200,000 pre-school age children a year. Studies have demonstrated that, in the long run, periodic vitamin A supplementation reduces deaths among pre-school age children in developing countries by about 25%. On the recommendation of the World Health Organization and UNICEF, more than one hundred countries worldwide now have implemented programs that give vitamin A to children. More than two million lives have been saved through these programs. With wider implementation and coverage of vitamin

A supplementation, an estimated six hundred thousand or more lives could be saved each year in developing countries.

Vitamin A public health efforts are generally seen as a success, but they have sometimes been stopped short by insurmountable obstacles – politics, cultural conflicts, bogus science, and commercial agendas. I detail many of these in this book. Vitamin A is not the only public health effort to have faced obstruction. In 2002, for example, religious clerics in northern Nigeria effectively opposed delivery of oral polio vaccine. The holy men saw the vaccination program as a plot to harm children. No surprise, large outbreaks of paralytic polio followed the halt of the program in 2003.

Nor are such impediments to improving wellbeing through public health programs limited to the developing world. Despite definitive evidence that fluoridation of water promotes oral health and reduces cavities in the general population, the Board of County Commissioners of Pinellas County, Florida, gave in to critics of so-called Big Brother government and voted in 2011 to end fluoridation of the water supply.

The incidence of whooping cough, which was common through World War II, was largely arrested thanks to the routine administration of three-pronged DPT (diphtheria, pertussis, tetanus) vaccinations – until recently, that is. Whooping cough is currently in resurgence in the United States, in part because of parents' refusal to have their children immunized.

This book could not have been written two decades ago. Only in the last twenty years has it been possible to define vitamin A deficiency as a nutritionally acquired immune deficiency syndrome. The new scientific certainties about vitamin A make possible a new historical interpretation. Whereas as recently as the early 1980s, a pediatrician might have said, 'This child is about to die from measles' or another infection, today we can correctly assign blame to an underlying culprit and take action against it. Likewise, an obstetrician might have closed the books on a lost patient, saying, 'Died of puerperal sepsis (childbed fever)'. Adequate intake of vitamin A, either before the onset of disease or once the patient was sick, may have ruled out both scenarios.

I began research for this book the year after my initiation in public health fieldwork, with questions about the seminal clinical observations of vitamin A deficiency made during the nineteenth century in such places as Paris, Bordeaux, and Lisbon. Army and navy medical records attest to the pervasive problem of vitamin A deficiency among soldiers and sailors. People by the millions, young and old, perished as a result of vitamin A deficiency. It took nearly two hundred years to understand what vitamin A was, that it could prevent or cure many diverse and deadly diseases, and how it did so.

There was no 'eureka!' moment of discovery in the quest to understand vitamin A. Nor was there one towering genius to could lay claim to understanding vitamin A and its biochemical powers. In this book, I attempt to reconstruct the twisted, broken path toward understanding vitamin A and toward introducing the men and women who, together, lifted the dark shadow that condemned to death the victims of vitamin A deficiency.

The preparation of this book was greatly facilitated by the superb assistance and expertise of the staff of the National Library of Medicine, especially Stephen Greenberg, Elizabeth Tunis, Kenneth Niles, Crystal Smith, and Khoi Le. I thank the staff of the Bibliothèque Nationale de France, the Österreichische Nationalbibliothek, the Caird Library at the National Maritime Museum in Greenwich, the National Archives at Kew, the Bibliothèque de l'Académie Nationale de Médecine, the University of Cambridge Library, the Bodleian Library at the University of Oxford, the Forbes Mellon Library at Clare College, the Library of the Royal College of Surgeons of England, the Rugby School Museum, the St. George's Medical School Library, the Kenneth Spencer Research Library, University of Kansas Libraries, the Wellcome Institute for the History of Medicine, Contemporary Medical Archives Centre, the Yale University Library, Manuscripts and Archives, and the Archives départmentales de la Gironde in Bordeaux. I also owe appreciation to Jean François Girardot of the Bibliothèque de Médecine, Université Henri Poincaré, Nancy, France, Florence Greffe of the Archives of the Académie des sciences – Institut de France, David Null at the Steenbock Library, University of Wisconsin – Madison Archives, John Hessler of the Geography and Map Division, Library of Congress, and Vickie Bomba-Lewandoski of the Connecticut Agricultural Experiment Station. Joanne Katz kindly shared her wealth of material from the Albay Child Health Project. I thank Kate Burns, Alfred Sommer, Omar Dary, Steve LeClerq, and Keith West Jr. for sharing their perspectives on vitamin A research and programs. My colleagues Martin Bloem, Klaus Kraemer, and Saskia de Pee provided valuable insight during the early formulation of the book. Kai Sun conducted the data analyses of Corry Mann's milk studies in children and of night blindness and infectious diseases in white and black troops during the US Civil War.

Rita Costa-Gomes provided invaluable assistance with works from the French, Portuguese, Spanish, and Italian literature. I thank Satoru Yamamoto, Kelly Barry and Christopher Wild, Anna Berchidskaia, and Michael Stern, respectively, for their translations of papers from the Japanese, German, Russian, and Dutch scientific literature. I am grateful to Kenneth Carpenter, Thomas Cohen, Sidney Mintz, and Adrianne Bendich for review of the early version of the book manuscript. Finally, I owe great thanks to Johanna Zacharias for her guidance, encouragement, and expert pre-submission editing which truly helped to bring this project to fruition.

Richard D. Semba

Glossary

Many terms used in this book occur only in a scientific/medical context; the brief definitions given here may be useful to the reader. Certain words that are familiar in nonscientific usage acquire distinct, specific meanings when used in a scientific/medical context (e.g. control, describe, synthesis, wasting). Their scientific/medical meanings are given here, along with many other frequently used scientific/medical terms.

Accessory food factors – an early twentieth-century term for the essential food components that became known, first, as 'vitamines' and, then, as 'vitamins'.

Adequate Intake (AI) – as determined by the Food and Nutrition Board of the Institute of Medicine (US), a recommended average daily nutrient intake level based on observations or estimates of apparently healthy people. AI is used when an RDA cannot be determined.

Albumin – a simple form of water-soluble protein such as that present in egg white, milk, and blood serum.

Amine – a group of organic compounds that are derivatives of ammonia.

Amino acid – a compound that contains carbon, oxygen, hydrogen, and nitrogen and is a 'building block' or basic unit that joins with other amino acids to form proteins.

Atropine – a substance that will dilate the pupils if instilled into the eyes.

Belladonna – an atropine-containing drug prepared from the deadly nightshade plant.

Basal – a diet that contains the caloric content to meet basic needs.

Beriberi – a nutritional deficiency caused by lack of thiamin and characterized by an array of clinical findings including loss of sensation in the extremities, paralysis of wrists and feet, burning sensation in the legs or toes, muscle atrophy and weakness, enlargement of the heart, and heart failure.

Bitot's spot – named for French physician Pierre Bitot (1822–1888), a raised, foamy or pearly-appearing patch of abnormal tissue arising on the surface of the conjunctiva and considered specific to the diagnosis of vitamin A deficiency.

Bran – the outer coats of cereal grains that are rich in B complex vitamins such as thiamin.

Calorie – a unit of heat energy mostly used to define the amount of energy in foods.

Carbohydrate – a group of organic compounds that includes starches, sugars, celluloses, and gums.

Carbonic acid – a weak acid present in solutions of carbon dioxide dissolved in water.

Butterfat – the natural fat in milk from which butter is made; butterfat is a rich source of vitamin A.

Buttermilk – the liquid remaining after butter has been separated from milk or cream; buttermilk is devoid of vitamin A.

Carotene – an orange-yellow to red pigment present in carrots, mangoes, papaya, and dark green leafy vegetables and a dietary precursor to vitamin A.

Carotenoids – a class of yellow to red pigments present in plants and animals.

Casein – the main protein in milk and cheese.

Case-fatality rate – the ratio of the number of deaths to the number of people with a given condition, for example, the case fatality rate of measles was 8% or 80 deaths per 1,000 children with measles.

Cod liver oil – an oil extracted from the liver of the cod fish; cod liver oil is a rich source of vitamin A.

Conjunctiva – the mucous membrane that lines the inner surfaces of the eyelids and the exposed surface of the eyeball.

Cornea – the transparent dome-shaped tissue that forms the front of the eye.

Contagious – of infectious diseases, spread person-to-person by direct or indirect contact.

Crystallize – to cause the formation of crystals.

Deficiency – a lack or shortage, for example, vitamin D deficiency.

Diarrhea, diarrheal disease – a condition characterized by frequent, loose, watery stools. Among infants and children, it is usually caused by harmful viruses or microorganisms; it is also especially severe in persons with underlying vitamin A deficiency.

Describe – in science, to identify.

Dementia – a deterioration of mental abilities such as memory, concentration, and judgment.

Dermatitis – inflammation of the skin; a skin rash.

Diet – the kinds of food that are regularly consumed.

Dysentery – an especially severe infectious form of diarrhea that is accompanied by blood and mucus in the stool.

Fat – compounds composed of glycerol and fatty acids that constitute the body's main energy storage.

Fibrin – an insoluble protein that forms during the clotting of blood.

Epidemic – the widespread occurrence of a disease or condition in a community or group at the same time.

Epidemiology – a branch of medicine that deals with the patterns, causes, and control of disease in populations.

Estimated Average Requirement (EAR) – as determined by Food and Nutrition Board of the Institute of Medicine (US), the average daily nutrient intake level required to maintain good health in about half of healthy people.

Gelatin – a colorless and odorless substance obtained by boiling the skin, bones, and tendons of animals in water.

Germ/germ theory – micro-organisms; the germ theory states that microorganisms are the cause of many diseases.

Hemeralopia – a term used to describe the impaired ability to see at night, i.e., night blindness.

Infectious – caused by a harmful micro-organism and transmitted through the environment.

Iris – the colored ring-shaped membrane of the eye that is located between the cornea and the lens of the eye.

Keratomalacia – a softening or melting of the cornea that occurs in the most advanced stage of vitamin A deficiency and usually results in blindness.

Lesion – a localized area of disease in a tissue.

Lactose – a sugar that is present in milk.

Lipid – a large group of organic compounds that are insoluble in water, oily in consistency, and includes fats, oils, waxes, sterols, and triglycerides.

Malnutrition – a condition that occurs when the body does not get the right amount of vitamins, minerals, or nutrients for optimal health.

Mineral – an inorganic element such as iron, calcium, potassium, sodium, or zinc that is essential to human, animal, and plant nutrition.

Niacin – a B complex vitamin found in meat, wheat germ (see Bran, above), and dairy products and is essential for nerve and digestive function.

Night blindness – impaired or no ability to see at night that, today, serves as a clinical indicator of vitamin A deficiency. Also referred to as hemeralopia or nyctalopia. (q.v.)

Nitrogen, nitrogenous – the nonmetallic element nitrogen that makes up nearly four-fifths of air and is also contained in proteins; compounds containing nitrogen.

Nyctalopia – a term used to describe the impaired ability to see at night, i.e., night blindness.

Nutritive – nutritious or nourishing.

Ophthalmia – an inflammation of the eye.

Ophthalmoscope – an instrument for viewing the interior of the eye, especially the retina.

Pasteurization – partial sterilization (killing of micro-organisms) of food by heating.

Pellagra – a nutritional deficiency caused by lack of niacin and characterized by dermatitis, diarrhea, and mental disturbances.

Polyneuritis – a term used to describe experimental beriberi induced in birds by a diet lacking in thiamin.

Protein – a large group of organic compounds that are essential constituents of living cells and consist of long chains of amino acids.

Pupil – the constricting/dilating opening at the center of the iris of the eye through which light passes to the retina.

Recommended Dietary Allowance (RDA) – as determined by Food and Nutrition Board of the Institute of Medicine (US), the daily amount of a specific nutrient that is required to maintain good health in practically all healthy people.

Retina – the light-sensitive membrane that lines the inside posterior wall of the eyeball and receives visual images, which are transmitted to the brain via the optic nerve.

Retinol – a chemical term for vitamin A.

Retinol Activity Equivalent (RAE) – as determined by Food and Nutrition Board of the Institute of Medicine (US), 1 microgram RAE is equivalent to 1 microgram of all-trans retinol, 2 micrograms of supplemental all-trans-beta-carotene, 12 micrograms of dietary all-trans-beta-carotene, or 24 micrograms of other dietary provitamin A carotenoids.

Rods and cones – the light-detecting cells in the retina that allow night and day vision, respectively.

Rickets – a nutritional deficiency in children caused by lack of vitamin D and characterized by weak, soft bones and impaired growth.

Rhodopsin – a light-sensitive pigment in the rods of the retina that converts light energy into a nerve signal.

Scurvy – a nutritional deficiency caused by lack of vitamin C and characterized by spongy, bleeding gums; a blotchy pattern of bleeding under the skin; weakness, and joint pain.

Spanish fly – a preparation made from dried, crushed blister beetles that causes blistering when applied to the skin.

Starch – a carbohydrate that is the main source of energy in plants, of which the most familiar sources in are rice, potatoes, and wheat.

Stunting – a condition of shortened stature caused by chronic malnutrition in children.

Syndrome – a group of signs and symptoms that together characterize a disease or abnormal condition.

Synthesize – in chemistry, to combine different chemical constituents to make a specific compound.

Thiamin – a B complex vitamin that is found in nuts, legumes, and whole grains and is essential for nerve function and metabolism.

Tryptophan – an amino acid that is essential in the diet because it cannot be synthesized by the body.

Ulcer, ulceration – an open sore on an external or internal surface of the body.

Vesicatory – a substance that causes blistering when applied to the skin.

Vitamin A – a fat-soluble vitamin found in liver, egg yolk, butter, and whole milk that is essential for normal growth, immunity, and vision.

Vitamin B – see niacin, thiamin

Vitamin C – a water-soluble vitamin, also known as ascorbic acid, that is found in fruit and vegetables and is essential for maintaining bones, teeth, and blood.

Vitamin D – a fat-soluble vitamin found in milk, fish, and eggs (also generated in the skin through direct sunlight exposure) that is essential for normal growth and development.

Wasting – a condition of abnormal thinness that is often associated with acute starvation or severe disease.

Xerophthalmia – a term that describes any one or more of the clinical findings of the eye with vitamin A deficiency: night blindness, Bitot's spots, corneal xerosis, corneal ulceration, keratomalacia, or corneal scarring.

Xerosis – a condition affecting the cornea in which the epithelium is altered with vitamin A deficiency and takes on a dry, glazed, whitish appearance.

Chapter 1

Vitamin A Deficiency in Nineteenth Century Naval Medicine

Ironically, technology, more than medical science, brought about a decline in the nineteenth century in the recorded incidence of night blindness at sea. The faster a ship could accomplish its transoceanic mission – that is, in weeks rather than months – the less uninterrupted time its sailors had to live under shipboard conditions, including on inadequate rations. Progress in marine propulsion thus translated into shortened periods of insufficient vitamin A in sailors' diets.

Technological advances revolutionized transoceanic travel in the late-1700s and the 1800s. With the introduction of the coal-fired steam engine for marine propulsion, the result of efforts by English, French, and American engineers, motor-propelled ships plied the oceans alongside vessels still reliant on venerable means, wind and sail. The main difference between the two was speed. A steam-powered ship with a paddlewheel – the first motorized means of marine propulsion – could complete a long journey in a fraction of the time that a sailing vessel required. In spring 1838, the British *Great Western* set a record by crossing from Bristol to New York in fifteen days. To do so required an average cruising speed of 8.2 knots (the equivalent of 9.4 miles per hour) [1]. This was nearly double the speed of that of an average sailing ship.

Commercial shippers were the first to seize the advantages of steam-powered propulsion, and the paddlewheel became the mainstay of civilian fleets by 1850. Navies, meanwhile, had to await further technological progress. Though faster and nimbler than sailing ships, and far less vulnerable to foul weather and turbulent waters, steam-driven paddlewheels had overwhelming drawbacks.

The wheels themselves, being mounted on a ship's sides, made easy targets for enemy fire. Moreover, the wheels, the boilers, and the coal fuel all occupied a significant portion of a craft's internal space, encroaching seriously on the room needed for artillery [2]. A steam-propelled paddlewheel armed with only a dozen guns always faced the prospect of confronting a wind-powered sailing ship's one-hundred and twenty guns. The technological advance that finally put steam power ahead of wind and sails for naval navigation was the screw propeller. Because the optimal placement of the screw propeller was the stern, it freed up the broadsides for artillery. And being mounted below the water line, it made a difficult target for enemy fire.

But until the screw became the primary means of propelling military vessels, navy sailors continued to have to withstand months-long periods at sea and to bear the attendant health hazards. Naval records, not the logs of merchant ships, therefore dominate the history of the illnesses that beset sailors on very long voyages. Navy crewmen, far more than merchant mariners, suffered the diseases caused by inadequate, ill-balanced diets. And it was the men on naval ships who challenged the physicians on board, who tried to understand and cure what ailed the sailors. Much of what is known about the nutrition-related diseases that affected sailors in the nineteenth century comes from the journals, diaries, and official records of those navy doctors.

Night Blindness at Sea

In late October 1860, the French warship *La Cornélie* set sail from the port of Toulon. East of Marseilles on the Mediterranean coast, Toulon was one of France's key departure points for building and defending the empire. *La Cornélie*, a sleek, three-masted corvette, was making her maiden voyage on that autumn day. The sun glinted off her fresh paint and polished brass, contrasting with the dull iron of her twenty-two cannons. *La Cornélie* was bound for the South Pacific.

A suitably seasoned crew would guide *La Cornélie*, with topmen Jacques Plée and Louis-Marie Stéphan to tend the sails and rigging, and Jean Denon in charge of the guns. Even seaman Elie Morin, though only twenty-four, had sailed the South Pacific. The health of these four, plus another two hundred and fifty-three officers, supervisors, servants, and seamen, was under the care of Marie-Louis-Eugène Chaussonnet, a physician in the employ of the French Navy.

With a brisk wind filling her sails, *La Cornélie* headed south into deep Mediterranean waters, then west past Gibraltar into the Atlantic Ocean. Sailing southwest, she reached South America, rounded Cape Horn, crossed the Pacific, and arrived within range of New Caledonia, Australia, and New Zealand – all within four untroubled months. Chaussonnet, who had been on the alert for complaints of loose teeth and spontaneous bleeding, noted in his journal with satisfaction that the crew was healthy, with 'not a single case of scurvy' on board [3].

By the end of 1862, *La Cornélie* had again traversed the Pacific and arrived at the coast of Chile, from which she sailed north toward Mexico. Suddenly, however, the tone of Chaussonnet's journal changed. Topman Plée came to the doctor complaining that he could no longer work at night: he could see perfectly well in daylight, but at night he was having difficulty seeing the rigging. And the problem was getting worse each night.

Identifying Plée's problem as acute night blindness, Chaussonnet followed a course that had been advocated by an American colleague and utilized by many other doctors in that era. He gave the topman five milligrams of strychnine to take by mouth each morning [4]. After four days the patient had no more symptoms.

But a month later Plée returned, this time with worse complaints: now, after twilight, he could see virtually nothing. Again Chaussonnet administered strychnine, and, in addition, he fumigated Plée's eyes mornings and evenings with ammonia water. One course to which Chaussonnet did not resort, although it had its advocates, was to induce vomiting with emetics. Nor did he use purgatives to cause intestinal evacuation [5]. After a week of the strychnine-and-ammonia regimen, Plée announced that his vision problem was cured and resumed his nighttime duties.

But other sailors began to appear with the same complaint, including Denon, Stéphan, and Morin. Soon the doctor was busy giving strychnine, fumigating the eyes, and applying medicines to the skin. Applied around the eye or to the nape of the neck, vesicatories – such as Spanish fly *[cantharis]* – caused skin blistering but were deemed beneficial because they caused irritation that would supposedly counter the disease [6]. Some of the afflicted seamen got better but then relapsed. Others simply got worse. Clearly, an epidemic was making its way through the crew of *La Cornélie*.

Chaussonnet had no previous experience with the disease he confronted, although its symptoms fit perfectly with the night blindness (*hemeralopia*, see textbox 1–1) described in an 1856 treatise by Jean-Baptiste Fonssagrives, a professor of medicine at Brest [7]:

> The nocturnal blindness is at first partial, the patient is enabled to see objects a short time after sunset, and perhaps will be able to see a little by clear moonlight. At this period of the complaint he is capable of seeing distinctly by bright candle-light. The nocturnal sight, however, becomes daily more impaired and imperfect, and after a few days the patient is unable to discriminate the largest objects after sunset or by moonlight; he gropes his way like a blind man, stumbles against any person or thing placed in his footsteps, and finally, after a longer lapse of time, he cannot perceive any object distinctly, by the brightest candle-light.

Plainly, the treatments with strychnine, fumigations, and vesicatories were not working. Many men returned after a few days with relapses or complaints of no relief at all. Chaussonnet therefore decided to change his approach and resort to a radical treatment advocated by a colleague in the army. This course entailed shutting a patient for at least a few days in a *cabinet ténébreux*, a dark closet [8]. From March to August 1863, forty of *La Cornélie*'s sailors spent time in the *cabinet ténébreux* before their vision returned – some, nearly two weeks in total darkness and one, a full month.

Textbox 1–1. Differentiating night blindness and naming it

Night blindness has been recognized in the West since antiquity and identified by many different terms. Aulus Cornelius Celsus, a first-century Roman scholar, called it *inbecillitas oculorum* (weakness of the eyes) [9]. The seventh-century Byzantine Greek physician Paulus Aegineta referred to it as *nyctalopia* (night blindness) [10]. In early modern history, the French surgeon Ambroise Paré too called it *nyctalopia*, while his follower Jacques Guilleaumeau wrote of *vespertina caecitudo* (evening blindness) [11]. The Dutch clinician Hermann Boerhaave referred to it as *visus diurnus* (sight by day), while in France, François Boissier de la Croix de Sauvages called it *amblyopia crepuscularis* (lazy eye of the dawn) [12]. In the Arabic-speaking world, meanwhile, medical scholars also applied diverse terms, including *shebkeret*.

By the mid-nineteenth century, nighttime vision problems were recognized as falling into two categories. The first, now termed *retinitis pigmentosa*, is congenital but rare [13]. A hereditary disease, it occurs mostly in families. It begins with moderate symptoms, mainly poor vision in low light and loss of peripheral vision, and worsens over time. Examination of the retina with an ophthalmoscope usually finds changes of pigmentation and narrowing of the blood vessels. No effective treatment for *retinitis pigmentosa* has yet been found.

The other night blindness can be severe in the early stages but is rarely permanent. In the nineteenth century it was often, but not universally, referred to as *hemeralopia* – an irony, since the Greek *hemera* means light. Until the cause was identified, it occurred in epidemics such as the one *La Cornélie* experienced. Usually acute and often transient, it afflicted several or many subjects living under the same, extreme conditions, such as on shipboard, in an army battalion, or in a prison.

The terminology in use today remains something of a muddle. Both *hemeralopia*, literally meaning difficulty seeing in bright light, and *nyctalopia*, referring to vision problems in low light, are commonly used interchangeably for night blindness. The conditions defined by the two terms are in fact each other's opposites. The mixup goes back to the writings of ancient Greek and Roman scientists and other physicians, and it has never has been definitively straightened out. The term more widely used throughout the eighteenth and nineteenth centuries, however, was *hemeralopia*, though nyctalopia can still be heard to mean the same condition [14].

This course of action, although ultimately effective, nonetheless frustrated Chaussonnet. It was slow to take effect, and, moreover, he did know not why it worked. Nor could he tell why only certain members of *La Cornélie*'s crew suffered from the disease. He looked for revealing patterns, pondering the common and divergent characteristics of the men who were and were not affected by night blindness. He arrived at a keen observation: none of the ship's fifteen officers, eight supervisors, eleven servants, was affected. What was making these three groups immune, while the general seamen were susceptible? Did an explanation lie in the conditions under which they performed their duties? The topmen, working the sails, spent their time high up in the rigging – that is, in open air and bright sunlight. The gunners, in contrast, worked mostly below deck in the half-light of crowded galleys. The ordinary seamen worked at various duties both above and below deck. Exposure to bright light versus darkness, and to wind versus shelter, seemed to Chaussonnet not to be causal factors. Could the seamen's susceptibility have anything to do with age or where they came from? Most of the affected seamen were between twenty and twenty-five years old and came from the west and northwest of France. As with the men's duties, Chaussonnet again found no link between age and birthplace, and susceptibility. He noted his frustration in his journal: enclosure in complete darkness '. . .is the only treatment to cure night blindness for most of the cases, and without it the disease cannot be cured, but it takes a lot of time: six, eight, ten, fifteen, and twenty days' [15].

Chaussonnet's discouragement echoed that of an army colleague, who spoke for the medical profession:

Night blindness is a 'strange' disease! There is not a single author who, in addressing this subject, does not make this remark, a sort of formulaic and obliged reference which has become today a stereotype of medical language, repeated incessantly by one or another. Nevertheless, what does it all mean? 'Strange', according to our dictionaries, means bizarre, fantasmagoric, capricious, extravagant; but can nature, in its manifestations, be considered extravagant, capricious, fantasmagoric, or even bizarre? Isn't night blindness a disease and, as such, a manifestation of nature, therefore included in the domain of science, as any other natural phenomenon? How can we say of any disease that it is 'strange'? [16].

Some physicians voiced the skepticism that certain seamen pronounced cured in fact were not. After days or weeks with no work to do and confinement in total obscurity to endure, some of the men merely feigned improved night vision so they could be released from the *cabinet ténébreux*. In any case, no resolution to the night blindness quandary was found while the epidemic aboard *La Cornélie* ran its course. The ship returned to Toulon in spring 1864, with her crew in much the same condition as it had been eighteen months before.

Chaussonnet's observations of the distinctions between men aboard *La Cornélie* who did and did not report night blindness seem not to have led anywhere, at least not immediately. The puzzle persisted, and other explanations abounded. One that was common among sailors was that sleeping on deck exposed to moonlight and humidity caused night blindness [17]. One physician's counter to this notion was that, were it correct, nearly all seamen would have night blindness because '. . .everyone who has sailed on state vessels knows that it would be almost impossible to prevent the sailors from falling asleep on deck during night duty, given the fact that the necessary labor requires the active participation of only a few of the them, and given that the peaceful state of the sea, the beauty of the sky, and the gentle rocking of the waves are such a strong invitation to sleep' [18].

Others argued that night blindness was a tropical malady caused by warm weather [19]. As boats left the tropics and entered temperate climates, however, the number of sailors with night blindness did not diminish [20].

Homesickness was another hazard of long voyages held responsible, and *hemeralopia nostalgique* (homesickness night blindness) was implicated because sailors who were affected sometimes were depressed to the point of wasting away [21].

In the French navy, sailors aboard warships stood watch on deck for six hours – *la grande bordée* – during which they could be exposed to continuous, direct light from the sun and bright reflections off the sea, the deck, the sails, plus the metal fittings and objects that were 'much too highly polished and shining these days' [22]. Some naval physicians therefore subscribed to the belief that so long an exposure brilliant light was causing night blindness [23]. Too much light, the theory held, led to 'an exhausted condition of the retina' [24].

To some physicians – and probably clergymen – night blindness was less a medical than a moral condition, specifically, an affliction of onanists. In 1841, a Dublin physician identified the 'sinful' and 'long continued indulgence of the most morbid sexual propensities, such as the constant usage of artificial means of arousing exhausted passions' as the cause of night blindness in six patients. Stopping 'the improper use of the genitals', he asserted, could prevent the condition: applying a silver nitrate solution to the glans penis could be a 'very effectual' means to avert further cases of night blindness [25]. Ascribing night blindness to masturbation may have seemed reasonable at the time, in light of the belief that healthy young men who, as phrased in Genesis, 'spilled their seed,' were poisoning their bodies, impairing their digestion, and generally wasting away [26]. After all, a noted Swiss physician and authority on masturbation had argued that the loss of one ounce of seminal fluid was equivalent to the loss of forty ounces of blood [27]. One doctor's cure for masturbation-caused night blindness was 'the discontinuance of the lamentable vice' combined with a 'generous diet' [28].

Whatever its cause, an epidemic of night blindness among crew members could disable a ship and cause a mission to be aborted. The French frigate, L'Andromède, setting out to sail the entire Pacific rim, was forced to halt its mission at the coast of South America because so many crew members reported sick with night blindness. The ship physician's journal reported:

> . . . three quarters of the crew were afflicted, and I could exempt only the most seriously affected crew members from night duty, thus one would often find on deck men who had trouble getting around on dark nights and others who, when steering, could only see the illuminated compass but could not see the sails. These sailors had so many relapses that I no longer counted on anything but the rain and their return to Brittany to effect a complete cure [29].

Night Blindness Linked to Other Diseases of Malnutrition

In the early nineteenth century, physicians reported that night blindness was associated with increased mortality. In 1819, the British naval surgeon Andrew Simpson described two cases of young sailors with night blindness who died from infections. The first died nineteen days after a bout of severe diarrhea; the second, with respiratory problems [30]. Simpson warned that medical practitioners should pay attention to night blindness, because other diseases combined with the night blindness to produce 'a fatal termination'.

The association of night blindness and diarrheal disease was also well recognized [31]. The Italian ophthalmologist Antonio Scarpa, for example, considered intestinal problems to be commonly associated with night blindness [32]. The chief surgeon of the French naval station in the Antilles, M. Barat, noted that the worse cases of night blindness were found in sailors who were suffering with diarrheal disease [33]. The outbreak of night blindness on the Prussian ship *Arcona* in 1861 was associated with dysentery and chronic diarrhea among the sailors, and many sailors died. Eitner, a naval physician, described xerosis of the cornea, an advanced eye lesion of vitamin A deficiency, in one sailor who died of chronic dysentery [34]. Something missing from the food – vitamin A – likely contributed to the great number of deaths among common seamen from diarrheal disease and other infectious diseases.

Vitamin A deficiency probably played a part, too, in the high death rate on British ships conveying convicted prisoners to Australia – a voyage that could take a sailing ship two hundred days in the eighteenth and early-nineteenth centuries. Of course, many of the involuntary passengers began the journey with their health severely compromised: prison diets usually provided little or no vitamin A. For example, in England and Wales in the mid-nineteenth century, the regulation prison diet consisted of oatmeal gruel, bread, cooked meat, potatoes, soup, molasses, and cocoa [35]. As one naval physician noted:

> In the convict ships proceeding to Australia, both scurvy and night blindness have frequently made their appearance, but the latter often escapes notice in consequence of the prisoners being sent down into prison either at or a little after sunset. Aboard the *Marquis of Hastings*, which conveyed prisoners to Hobart Town in 1841, many cases of scurvy occurred, and there were ten of night-blindness, which presented no other symptoms of scorbutic disease [36].

Vitamin A deficiency probably also contributed to increased shipboard injuries and deaths through trauma resulting from accidents and falls [37].

Records of the vigorous transoceanic maritime activity of the nineteenth century – stimulated by the building and defending of empires, civilian travel, and trading in slaves as well as legitimate merchandise – attest to widespread occurrence of *hemeralopia*. That, in turn, drew extensive medical attention to the phenomenon. More than one hundred reports of night blindness on ocean voyages were published in the nineteenth century alone. A British naval surgeon in the West Indies station, Sir John Forbes (1787–1861), for example noted that night blindness was common. 'I have known it to exist in a proportion greater than one in twenty, and have been informed by surgeons and other officers on that station that they have seen it prevailing in ships to a much greater extent' [38]. As was the case with *La Cornélie*, the condition was most common in men who put to sea for many consecutive months [39].

For the physical well-being of the crewmen, the vigorous slave trading conducted under several nations' flags was the most pernicious of all maritime undertakings. Portugal and Spain, for example, had doubled their trading in slaves from West Africa to the New World to 135,000 a year by 1840. Great Britain, before eventually abolishing slavery along with most other European nations, took the high moral ground, and based a small fleet at the West African squadron to intercept illegal slave trading. British ships patrolled the area from Cape Verde in the north to Cape Negro south of the Equator. Alexander Bryson, a physician and British naval officer in charge of statistical reporting, declared this fleet's work to be 'the most disagreeable, arduous, and unhealthy service that falls to the lot of British officers and seamen' [40].

Bryson described slavers' methods: 'To avoid observation slaves were seldom embarked till the dusk of the evening, and this, which seldom occupied more than an hour or two, according to the nature of the bar or surf, having been effected, the vessel immediately made all sail, and endeavored to gain an offing beyond the cruiser, if possible, before daybreak' [41]. He also described how the men aboard the cruisers flying the Union Jack conducted the job of surveillance. Six men – two at the bow, two amidships – were assigned '. . .to scan the horizon with a night glass', – that is, until night blindness struck.

Bryson noted the effects aboard the British brigantine *Griffon* in 1851: '. . .out of about fifty white men twenty-two were affected, and immediately after the sun went down, they had to be led about on the upper deck, in a helpless state of blindness. There was now just cause for alarm, as the vessel with so many men unfit for night duty, was hardly a match for any of the well-armed slavers so common on the coast at that period' [42].

The crews of West Africa squadron ships faced yellow fever, diarrhea and dysentery, malaria, scurvy, and more. Bryson noted the squadron's exceptional annual mortality between 1825 and 1845 from infectious diseases alone: 54.4 deaths per 1,000 men of the mean force employed. This mortality rate was much higher than that found in the other squadrons – South America, 7.7 per 1,000; the Mediterranean, 9.3 per 1,000; Home (England), 9.8 per 1,000; the East Indies, 15.1 per 1,000; and the West Indies, 18.1 per 1,000 [43].

The lurid conditions in the British naval hospital on the island of Fernando Pó off Cameroon acquired a widespread reputation. Sailors made a joke of the situation: the standing orders were, 'Gang No. 1 to be employed digging graves as usual. Gang No. 2 making coffins until further orders' [44].

Night blindness sometimes occurred on long voyages during outbreaks of scurvy, leading some physicians to conclude as early as the eighteenth century that, besides the bleeding gums, red blotches on the skin, and loss of teeth, it too was a characteristic of scurvy [45]. In his 1785 treatise, *Observations on the diseases incident to seamen,* the British naval physician Sir Gilbert Blane cited night blindness as a symptom of scurvy [46]. On the French frigate *La Belle-Poule* off Madagascar, night blindness affected one hundred and eighty seamen, of whom all but four also had signs of scurvy [47]. Scurvy and night blindness were also closely associated in large outbreaks, for example, among seventeen men on *Le Colbert* in the Gulf of Mexico in 1864, and among thirty-three men on *L'Embuscade* and seventy-five men on *L'Alceste* in the Pacific [48]. The night blindness that attacked many crew members of the British squadron during the Siege of Gibraltar was attributed to scurvy. Many physicians therefore surmised that night blindness was just one point in the constellation of signs and symptoms of scurvy.

Detailed logs from the 1857–1859 global circumnavigation of the Austro-Hungarian frigate *Novara* – a voyage of extraordinarily long duration – shed valuable scientific light on the different causes of night blindness and scurvy [49]. A distinguishing characteristic, observable most clearly over so long a period at sea, is the length of time each condition takes to develop. In brief, scurvy can appear within a span as short as four months at sea. Night blindness, in contrast, may not develop in sailors on the same mission for as long as a full year. The explanation lies in the foods available to the sailors and how their bodies could and could handle the vitamins in those foods (textbox 1–2).

Textbox 1–2. Scurvy versus night blindness: a lesson from the *Novara*

The voyage of the *Novara* provides a dramatic contrast of the time it takes to develop vitamin A deficiency versus vitamin C deficiency. Vitamin C (ascorbic acid) is essential to many important biological functions (e.g. the formation of the connective tissues' collagen and the synthesis of neurotransmitters; it is also an antioxidant). The body has no organ that can store vitamin C, however, so with no dietary intake of vitamin C, an adult can develop scurvy within just three or four months.

Vitamin A, in contrast, has a storage organ: the liver. When the dietary intake of vitamin A stops, a healthy adult with a recent history of adequate intake may have as much as a year's worth of vitamin A in reserve. Experiments conducted in the 1940s with volunteer subjects recruited from among Great Britain's conscientious objectors showed that concentrations of vitamin A in the blood start to drop in some subjects after about eight months. Moreover, most vitamin A-deprived subjects showed some abnormalities in night vision after several months [50].

In the case of *Novara*'s circumnavigation, cases of scurvy generally appeared during long legs of the voyage and not between the ship's many stops at ports where sources of vitamin C such as citrus fruits and potatoes were abundantly available. According to the *Novara*'s log, the ship put in at Madeira, Rio de Janeiro, Cape Town, Ceylon, Madras, Singapore, Shanghai, Sydney, Tahiti, Valparaiso, and Gibraltar [51].

Night blindness, however, appeared in the *Novara*'s towards the end of the circumnavigation, when the seamen's vitamin A levels were presumably at their lowest.

Night blindness and the other diet-related disorders with which it was associated were by no means limited to long ocean voyages. Pellagra, for instance, with its characteristic skin lesions, diarrhea, wasting, as well as neurological and psychiatric disturbances (medical students often refer to pellagra's '3-Ds' – dermatitis, diarrhea, and dementia), was sometimes associated also with night blindness. Seen far more often on land than at sea, pellagra was once common in rural France, Italy, Spain, and the southern United States. Its usual victims were peasants and sharecroppers whose diet was poor in dairy products and meat. As early as the 1780s, Italian physician Gaetano Strambio observed several peasants with both pellagra and night blindness [52]. Louis Billod, a French psychiatrist, attributed night blindness to the general debility and sun exposure found in patients with pellagra, although he also noted that night blindness could occur by itself [53]. In the 1860s, France's Théophile Roussel, a physician, legislator, and leading authority on pellagra, also noted an association between that disease and night blindness [54].

Malaria, too, was sometime linked to night blindness. The British naval surgeon Andrew Simpson described a sailor who developed night blindness of nearly two weeks' duration after recovering from a malaria attack [55]. French and Italian physicians, too, considered their malaria patients to be predisposed to night blindness [56]. Similarly, a Spanish physician observed night blindness to be present in regions where the malaria rate was high [57].

Hookworm infection (helminthiasis), too, was occasionally associated with night blindness [58]. The Italian ophthalmologist Antonio Scarpa considered treatments to expel worms also useful in treating patients with night blindness [59].

Diagnosis and the Search for a Cause

Until the development of the ophthalmoscope in 1850, a diagnosis of night blindness was usually made on the basis of a patient's complaint and on the external appearance of his eyes – specifically of the pupils. Night-blinded eyes have a distinctive feature: the openness of the pupils is inappropriate to the light level of the environment. In healthy eyes, the pupils react to bright light by constricting and to dark or dim conditions by dilating. Eyes affected by hemeralopia lack this responsiveness.

The pupils' reaction to light is under the control of the retina. A diseased retina can impede the normal action of the pupils, causing the pupils to respond to changes in light levels either sluggishly or not at all. Pupil size, therefore, was a clue but not the basis of a definitive diagnosis in the nineteenth century. Indeed, sailors were sometimes suspected of using this recognized clue as the basis for trickery. The suspicion arose that a would-be patient, seeking relief from his duties, might have instilled belladonna in his eyes, thus emulating one key sign that was associated with night blindness. The name of this drug, which would have been on board for the treatment of intestinal disorders, means beautiful woman for the very fact that it causes enlarged pupils that enhance the eyes' loveliness.

Whether or not sailors actually resorted to the belladonna ploy, it was made impossible by the widespread use of the ophthalmoscope. This revolutionary instrument enabled a physician to look through the pupil directly into the interior of the eye and at its back – that is, at the retina. By examining a retina with an ophthalmoscope, a

Table 1.1. Classification of night blindness according to Dr. Piriou in 1865

Grade	Criteria
1	Impossible to distinguish clearly the moon
2	Possible to distinguish the moon
3	Possible to distinguish the planets and Sirius
4	Possible to distinguish stars of primary magnitude, lunar light, and the phosphorescent glow of the sea
5	Possible to distinguish stars of secondary magnitude
6	Possible to distinguish stars of third magnitude and light of a clear atmosphere

physician can distinguish a case of retinitis pigmentosa from a case of hemeralopia (textbox 1–1) [60].

Before the widespread use of the ophthalmoscope, different methods were developed for diagnosing and grading the severity of night blindness. Great Britain's Andrew Simpson described a vision test in 1819 in which affected patients were asked at twilight to discern different-sized dots on white paper [61]. In 1865, French naval physician Piriou developed a six-level scale based on ability to see light sources in the sky and the sea (table 1.1) [62]. Jean Boudet disparaged both these methods and, in 1871, developed his own method based on the diameter of the pupil [63]. He pasted eleven black circles on white cardboard, with each circle larger in diameter by one millimeter than the previous; the smallest circle was two millimeters across, and the largest, twelve millimeters. The physician could place this pupillary scale next to the patient's eyes to gauge the degree of night blindness.

Textbox 1–3. A clever test for diagnosing a true case of night blindness

In the mid-19th century, the German ophthalmologist Alfred Graefe noted that, in a normal subject, if a glass prism were placed in front of one eye to displace the image, the patient would move both eyes to overcome the prism, fuse the image, and avoid double vision. If a subject with night blindness were situated in some degree of darkness, when a prism was placed in front of one eye, the patient's eye would not move to fuse the image, presumably because of his inability to see in the dim light [64]. Thus, Graefe advocated the use of this prism test to diagnose true cases of night blindness.

To prevent night blindness, of course, a cause had to be found. With the rate of night blindness so high among sailors, excessive exposure to bright light seemed an obvious culprit. Carl Stellwag von Carion, a Viennese ophthalmologist, declared definitively in 1867, 'The immediate cause of hemeralopia is over-dazzling of the eyes' [65]. Another observer noted in 1888, '. . .in the middle of the sea, in the tropics where the sky is always clear and the night dazzling, [night blindness] strikes caulkers, topmen, and all those whose duty renders them defenseless against its influence' (see textbox 1–4) [66].

Accordingly, one naval surgeon proposed the use of broad-brimmed straw hats for seamen [67]. Another advised the use of dark blue eyeglasses [68]. By the 1870s, when steam-powered vessels shared the seas with sailing ships, different precautions were advocated: on sailing ships, sailors could find refuge in the sails' shadow; on steamships, shade could be provided with an awning or tent on deck [69]. One physician proposed that blue curtains be suspended from the awnings when the ships were in the bright sun [70].

A far more economical method of prevention that was tried and declared highly successful involved patching the sailors' eyes, one at a time. Reporting in 1868 on a hellish entanglement at sea involving a British naval ship, a Spanish privateer, and an intruding pirate vessel, Nottidge MacNamara, Professor at Calcutta Medical College, wrote with considerable medical knowledge as well as practical sense:

> In a few days, at least half the crew were [sic] affected with nyctalopia. . . .
> This circumstance put me on devising some means of curing the people affected with night blindness, and I could think of none better than excluding the rays of the sun from one eye during the day, by placing a handkerchief over it; and I was pleased to find, on the succeeding night, that it completely answered the desired purpose, and that the patients could see perfectly well with the eye which had been covered during the day; so that, in future, each person so affected had one eye for day, and the other for night. It was amusing enough to see [a sailor] guarding, with tender care, his night eye from any of the slightest communication with the sun's rays, and occasionally changing the bandage, that each eye in turn might take a spell of night duty, it being found that guarding the eye for one day was sufficient to restore the tone of the optic nerve, a torpor of which and of the retina is supposed to be the proximate cause of the disease. I must question whether any purely medical treatment would have had so complete, and above all, so immediate an effect [71].

In the search for a cause, the question posed by *La Cornélie*'s physician Chaussonnet – Why do only certain groups of men on shipboard suffer from night blindness? – did receive some serious consideration. Pondering on why night blindness tends to attack common seamen and to spare officers, servants, and cabin boys, one physician felt that the lowly sailors' eyes were naturally predisposed to blinding when these men came topside into the sunshine from their dark quarters below: 'When eyes accustomed to this half-darkness are struck by the bright daylight on the upper deck, the cleaned metal objects shining in the sunlight, even by the sailors' white clothing, a blinding sensation occurs' [72].

Communal living, bad hygiene, and poor ventilation were also implicated as predisposing factors for night blindness among the common seamen [73].

Something Missing from the Food

In 1787, when the British First Fleet began transporting convicts to Australia, the weekly government-provided victuals for each crewmember that consisted of salt beef or salt

Table 1.2. Rations in the British Navy, 1740–1825[1]

Food	Sunday	Monday	Tuesday	Wednesday	Thursday	Friday	Saturday
Biscuits, lbs	1	1	1	1	1	1	1
Beer, gallons	1	1	1	1	1	1	1
Beef, lbs	0	0	2	0	0	0	2
Pork, lbs	1	0	0	0	1	0	0
Peas, pints	2	0	0	2	2	2	0
Oatmeal, pints	0	1	0	1	0	1	0
Butter, oz[2]	0	2	0	2	0	2	0
Cheese, oz[3]	0	4	0	4	0	4	0

[1] Both butter and cheese were removed from the rations in 1825.
[2] Olive oil substituted for butter on long voyages.
[3] Cheese was 'flet cheese', a low-quality skim-milk cheese from Suffolk.

pork, dried peas, oatmeal, biscuits, cheese (12 ounces), butter (6 ounces), and vinegar [74]. Olive oil, which contains no vitamin A, was substituted for butter because the latter went rancid on long voyages. The cheese was an inexpensive 'flet' cheese, made from skim milk and containing virtually no vitamin A. The rations were of course even worse for the convicts: males were allowed one third less than the seamen, and females were allowed two-thirds of what the male convicts' ration, or about half that of the crew.

Thomas Trotter, another physician of the Royal Navy, wrote in his 1799 *Medica nautica: an essay on the diseases of seamen*, that the seamen's diet should be a 'branch of medicine of the first importance'. He asserted that, '. . .if one half of the money expended on chests of medicine were laid out in the comforts of diet, much real advantage would be gained' [75]. Despite Trotter's insistence, the navy's regulation diet stayed the same for nearly a century (table 1.2) [76].

Finally, in 1825, the daily meat and beer allowances were increased, but cheese and butter, which were difficult to preserve, were removed [77]. With butter and cheese off the list throughout most, the diet of Britain's seamen was virtually devoid of vitamin A. Later modifications added chocolate, raisins, and preserved potatoes, but still no butter or cheese.

Great Britain's Alexander Bryson was in the vanguard of physicians to point to bad food as a cause of night blindness. He attributed hemeralopia, like scurvy, to the lack of fresh meat and vegetables, asserting that night blindness was 'entirely dependent on an improper or erroneous diet' [78]. France, at least, approached the matter with some sensitivity to the pleasure that food can give, and innovations in the diet of the French navy included the baking of fresh bread on board ship and the introduction of coffee [79].

In the 1840s, after the development of a technique for desiccating vegetables, a factory was established in Paris to produce dried vegetables. The French navy sought

Table 1.3. Rations in the French Navy, 1877

Type of food or drink							
Fresh bread (750 g)			or		Biscuit (550 g)		
beef conserve (200 g) with	beans (60 g) or	or	bacon (225 g) with	beans (60 g) or	or	fresh beef (300 g) with	green beans
	peas (60 g) or			peas (60 g) or			
	mixed dried vegetables (potatoes, broadbeans, cabbage) (18 g)			mixed dried vegetables (potatoes, broadbeans, cabbage) (18 g)			
On Fridays, in lieu of the above: beans (60 g) or sardines in oil (70 g) with beans (60 g)							
When in port: cheese (80 g)							
Dried beans (230 g) or peas (230 g)							
Bacon (80 g)							
Rice (80 g)							
Eau-de-vie (60 ml) or rum (60 ml) or tafia [West Indian rum] (60 ml)							
Wine (460 ml)							
Coffee (20 g)							
Brown sugar (25 g)							
Seasonings (chutney, olive oil, fats, mustard, pepper, salt, vinegar)							

to determine whether the vegetables could be used to increase the variety in the diet. The dried vegetables were tested on a voyage of *Le Caïman* to the Red Sea in 1853 and 1854, and met with some acceptance by the crew [80]. In 1856, discussion began about introducing dried vegetables in the regular rations of the French navy, and dried vegetables were introduced to some extent after 1861 [81].

Nevertheless, the French navy diet of 1877, like its British counterpart, remained grossly deficient in vitamin A (table 1.3) [82], and Spanish sailors fared little better (table 1.4) [83]. The French sailors' rations consisted mostly of salt-preserved meat and pork, bread or biscuits, dried peas, and dried vegetables. For vegetables, there were desiccated potatoes, broad beans, and cabbage (the last containing trace amounts of vitamin A). Olive oil, rather than butter, served as the basis for cooking. Cheese was offered primarily when the ships were in port.

In contrast to the navy, the English East India Company attempted to provide its ships with foods that were the 'very best in their kind, and with respect to the quantity allowed much exceed that in any service' [84]. The company's shipboard provisions

Table 1.4. Rations in the Spanish Navy, 1805

	Sunday	Monday	Tuesday	Wednesday	Thursday	Friday	Saturday
Biscuit, oz	18	18	21	18	18	18	18
Beef, oz	3	0	0	0	0	0	0
Pork, oz	2	4	2	4	4	2	4
Menestras, oz	5	5	2	5	5	3	5
Cheese, oz	0	0	0	0	0	5	0
Vinegar, quartillo	0	0	0.25	0	0	0	0
Oil, oz	0	0	2	0	2	0	0
Garlic (head)	0	0	2	0	0	0	0
Wine, quartillo	2	2	2	2	2	2	2
Salt, quartillo	0.001	0.001	0.001	0.001	0.001	0.001	0.001

consisted of ample amounts of salt beef and pork, stockfish, chickpeas, flour, peas, yams, brandy, or arrack – but, again, no substantial sources of vitamin A. Compared with seamen from Europe, the East India Company's lascars – seamen for hire born on the Indian subcontinent – were reputedly more susceptible to night blindness [85].

Textbox 1–4. Light, the eye, and vitamin A

The pigment rhodopsin, also sometimes called visual purple, is the essential substance for the eye's adaptations to light and dark conditions. Rhodopsin resides in the retina, specifically in the rod photoreceptors; it is not present in the retina's other photoreceptors, the cones. The cones and rods have complementary functions: to enable visual perception in light and dark conditions, respectively. Vision relies on the cones during the day. As the day gives way to night, responsibility for seeing passes from the cones to the rods.

The transition – that is, adaptation to darkness – is gradual and can take as long as a half hour to become complete [86]. During the dark adaptation process, the eyes transfer their reliance from the cones to the rods, both of which must be in good working order to enable both day and night vision.

Whereas darkness – partial or total – cannot damage or impede the function of the retina, too much light can, because excessive exposure to bright light bleaches the rhodopsin in the rods. In healthy eyes, however, the bleaching is not permanent. A person going from bright conditions to dark undergoes a period of partial or total blindness, which, ideally, lasts only until the rods have recovered their rhodopsin. Eyes lacking sufficient rhodopsin, however, make this recovery poorly.

The adequacy of rhodopsin is a function of food – specifically, of the amount of vitamin A in a person's regular diet. The less vitamin A a person ingests, the likelier he is to suffer from night blindness. Butter and cheese are two of the foods that boost a person's supply of vitamin A. When ships were at sea for long periods in the nineteenth century, the sailors who manned them had little or no butter or cheese in their diets for months on end.

While Great Britain's lowly seamen made do with inadequate and unappetizing rations, the officers enjoyed a substantially different diet. The captain and officers usually had their own cook and stocks of food and wine. In addition, many ships had livestock on board to provide fresh eggs, milk, and cheese. Commodore Anson had his own French cook on board the *Centurion* during his celebrated circumnavigation in the mid-eighteenth century. Aboard the *Prince George*, as she made her way from Spithead to New York, Admiral Robert Digby dined on roast duck, butter, potatoes, carrots and greens, and plum pudding [87]. But no one seems to have drawn any conclusions about who on board was and was not reporting sick with night blindness. The mysterious epidemics of night blindness, disease, and deaths slowly disappeared from naval records towards the end of the nineteenth century. As to their cause, further clues were come from investigations conducted far from the sea.

References

1 Fry, H. (1896) The history of North Atlantic steam navigation, with some account of early ships and shipowners. London, S. Low, Marston, p. 41.

2 Black, J. (2004) The British seaborne empire. New Haven and London, Yale University Press, p. 210.

3 Chaussonnet, M. L. E. (1870). De l'héméralopie aiguë; thèse pour le doctorat en médecine. Faculté de médecine de Paris, No. 63. Paris, A. Parent.

4 Chaussonnet (1870), p. 7; Mora, M. J. (1871) Hemeralopia – cura pela strychnina. O Correio Medico de Lisboa 1, 55–57; Chisholm, J. J. (1871) Nightblindness, of seven months duration, resisting the usual treatment, but promptly relieved by the hypodermic use of strychnia. Baltimore Medical Journal and Bulletin 2, 392–398.

5 Robert, A. (1846) Cas d'héméralopie chez un ouvrier travaillant dans une carrière de grés. Guérison. Remarques cliniques sur cette rare affection. Lancette Française 2nd series, 8, 242–243; Fonssagrives, J. B. (1852) Histoire médicale de la campagne de la frégate à vapeur l'Eldorado. (Station des côtes occidentales d'Afrique, années 1850–1851); thèse pour le doctorat en médecine. Faculté de médecine de Paris. No. 136. Paris, Rignoux, p. 53.

6 Bamfield, R. W. [sic] (1819) A practical essay on hemeralopia, or night-blindness, commonly called nyctalopia; as its effects [sic] seamen and others, in the East and West Indies, China, the Mediterranean, and all tropical climates; in which a successful method of curing the disease is detailed. Medico-Chirurgical Transactions 5, 32–66. [Author was Robert W. Bampfield, surgeon of the Royal Navy]; Guépin, A. (1858) Deux observations d'héméralopie recueillies au dispensaire ophtalmologique du Dr. Guépin, à Nantes. Annales d'Oculistique 39, 48–51.

7 Fonssagrives, J. B. (1856) Traité d'hygiène navale; ou de l'influence des conditions physiques et morales dans lesquelles l'homme de mer est appelé à vivre et des moyens de conserver sa santé. Paris, J. B. Ballière.

8 Netter, A. (1858) Du traitement de l'héméralopie par l'obscurité. L'Union Médicale 12, 450, 455–456.

9 Celsus, A. C. (1542) De re medica libri octo. Item Q. Sereni Liber de medicina. Q. Rhemnii Fannii Palaemonis De ponderibus & mensuris liber. Lugduni, Seb. Gryphium, p. 290.

10 Paulus Aegineta. (1538) Libri septem, quibus dextra medendi ratio ac via tam in diaetetico, quàm pharmaceutico & chirurgico genera compendiò cōtinetur, per Albanum Torinum, vito durensem partim recogniti, partim recens latinitate donati. Basileae [Per Balthasarem Lasium], 1538.

11 Paré, A. (1685) Les oeuvres d'Ambroise Paré, conseiller et premier chirurgien du roy. 13th edition. Lyon, Pierre Valfray; Guillemeau, J. (1587) A worthy treatise of the eyes, containing the knowledge and cure of one hundred and thirtene diseases, incident vnto them: first gathered & written in French, by Iacues Guillemeau, chyrurgion to the French King, and now translated into English, together with a profitable treatise of the scorbie; & another of the cancer by A.H. Also next to the treatise of the eies is adoiyned a work touching on the preseration of the sight, set forth by VV. Bailey, D. of Phisick. London: printed by Robert Waldergrove for Thomas Man and William Brome.

12 Boerhaave, H. (1750) De morbis oculorum. Gottingen, Vandenhoeck, p. 159; Boissier de la Croix de Sauvages, F. (1785) Nosologia methodica oculorum; or, a treatise on the diseases of the eyes. London, G. G. J. and J. Robinson, p. 260.

13 Richter, H. C. E. (1828) Dissertatio inauguralis medica exhibens tres hemeralopiae s. caecitatis nocturnae congenitae casus additis quibusdam adnotationibus hunc morbum in universum spectantibus. Jenae, Schlotteri; Cunier, F. (1838) Histoire d'une héméralopie héréditaire depuis deux siècles dans une famille de la commune de Vendémian, près de Montpellier. Annales de la Société de médecine de Gand 4, 385–395; Stiévenart. (1847) Note sur une héméralopie héréditaire. Annales d'Oculistique 18, 163–164.

14 Greenhill, W. H. (1880–1882). On the meaning of the word 'nyctalopia' and 'hemeralopia' with a critical examination of the use of these words in the ancient Greek and Latin authors. Ophthalmic Hospital Reports, London, 10, 284–292; Tweedy, J. (1880–1882) On the meaning of the words 'nyctalopia' and 'hemeralopia' as disclosed by an examination of the diseases described under these terms by the ancient and modern medical authors. Ophthalmic Hospital Reports, London, 10, 413–436.

15 Chaussonnet (1870), p. 44.

16 Netter, A. (1862–1863) Nouveau mémoire sur l'héméralopie épidémique et le traitement de cette maladie par les cabinets ténébreux. Gazette Médicale de Strasbourg 22, 164–171, 186–192; 23, 9–17, 21–27.

17 Forbes, J. (1811) Observations on tropical nyctalopia. Edinburgh Medical and Surgical Journal 7, 417–419; Audouit. (1855) De l'héméralopie, observée dans les voyages de circumnavigation. Archives d'Ophthalmologie 4, 80–106; Centervall, I. A. (1897) A case of moon-blindness. Boston Medical and Surgical Journal 137, 383.

18 Audouit (1855), p. 94.

19 Forbes (1811), p. 417; Baillie, H. (1816) Observations on the hemeralopia, or night blindness. Medico-Chirurgical Journal and Review 2, 179–182.

20 Audouit (1855), p. 93.

21 Coquerel, C. (1849) De la cécité nocturne. Thèse pour le doctorat en médecine. Faculté de médecine de Paris, no. 150. Paris, Rignoux.

22 Fleury, E. J. (1840) Note sur l'héméralopie épidémique. Gazette Médicale de Paris 8, 50–54.

23 Coquerel (1849), p. 15; Deval, C. (1858) Considérations pratiques sur les principales variétés de l'héméralopie et sur le traitement qui leur est applicable. Bulletin Général de Thérapeutique 55, 248–256, 303–308; Walton, H. (1866) The substance of a lecture on night-blindness. Symptoms – cause – pathology – results – treatment. Medical Times and Gazette (London) 1, 169; Stellwag von Carion, K. (1868) Treatise on the diseases of the eye, including the anatomy of the organ. New York, William Wood, p. 656; Fernandez-Caro, A. (1882) La hemeralopía. Boletin de Medicina Naval 5, 6–12.

24 Robinson, C. H. (1868) Remarks on night-blindness. Lancet 1, 683–684.

25 Cane, R. (1840–1841) Cases of 'night,' or 'moon blindness,' and of ordinary amaurosis, caused by onanism and inordinate venery; with remarks. Dublin Journal of Medical Science 18, 169–181; Banerjee, H. P. (1901) Night-blindness, its causation, varieties, peculiar symptoms of each variety and treatment. Indian Lancet 18, 937–940.

26 Gilbert, A. N. (1975) Doctor, patient, and onanist diseases in the nineteenth century. Journal of the History of Medicine and Allied Sciences 30, 217–234.

27 Tissot, S. A. D. (1764) L'onanisme. Dissertation sur les maladies produites par la masturbation. 3 edition. Lausanne, Marc Chapuis et compagnie.

28 Cane (1840–1841), p. 181.

29 Audouit (1855), p. 82.

30 Simpson, A. (1819) Observations on the hemeralopia; or, nocturnal blindness, with cases and practical illustrations. Glasgow, R. Chapman.

31 Guépratte, A. (1847) Héméralopie des pays chauds. Observations médicales recueillies à bord de la frégate l'Armide – Mission de Madagascar – 1846. Gazette Médicale de Montpellier 8, 6–7; Ouvrard, C. F. (1858) Quelques remarques sur l'héméralopie observée à bord du Lavoisier, pendant une campagne en Océanie. Thèse pour le doctorat en médecine. Faculté de médecine de Paris, no. 298. Paris, Rignoux, p. 19; Saurel, L. (1861) Traité de chirurgie navale. Paris, J. B. Baillière et fils, p. 428.

32 Bégin, Boisseau, Jourdan, Montgarny, Richard, Sanson, Dupuy. (1823) Dictionnaire des termes de médecine, chirurgie, art vétérinaire, pharmacie, histoire naturelle, botanique, physique, chimie, etc. Paris, Crevot, Béchet, Baillière, p. 244.

33 Audouit (1855), p. 86.

34 Eitner. (1863) Eine Epidemie von Hemeralopia, beobachtet auf Sr. Maj. Schoff Arcona während der ostasiatischen Expedition. Deutsche Klinik 15, 245–248.

35 Pereira, J. (1843) A treatise on food and diet: with observations on the dietetical regimen suited for disordered states of the digestive organs; and an account of the dietaries of some of the principal metropolitan and other establishments for paupers, lunatics, criminals, children, the sick, etc. New York, Fowler and Wells.

36 Bryson, A. (1859–1860) Night-blindness, in connexion with scurvy. Ophthalmic Hospital Reports, London 2, 40–43.

37 Fonssagrives (1856), p. 357; Saurel (1861), p. 431.

38 Forbes (1811).

39 Bulkeley, J., Cummins, J. (1797). A voyage to the South Seas, in the years 1741–1. Second edition, with additions. London, James Coutin, pp. 10–11; Blane, G. (1785). Observations on the diseases incident to seamen. London, Joseph Cooper, p. 22; MacNamara, N. C. (1868). A manual of the diseases of the eye. London, John Churchill & Son, p. 406.

40 Bryson, A. (1847) Report on the climate and principal diseases of the African station; compiled from documents in the office of the director-general of the medical department, and from other sources, in compliance with the directions of the right honorable the lords commissioners of the admiralty. London, William Clowes and Sons, p. 161.

41 Bryson (1847), p. 2.

42 Bryson (1859–1860).

43 Bryson (1847), p. 178.

44 Lloyd, C., Coulter, J. L. S. (1963) Medicine and the navy, 1200–1900. Volume IV. 1815–1900. Edinburgh and London, E & S Livingstone, p. 159.

45 Nicolls, G. A. (1855) Scurvy and hemeralopia [letter]. Medical Times and Gazettte London n.s. 2:96.

46 Blane (1785), p. 46.

47 Vaucel, A. (1891) Contribution à l'étude de l'étiologie de l'héméralopie épidémique. Thèse pour le doctorat en médecine. Faculté de médecine et de pharmacie de Bordeaux. Année 1890–91, No. 40. Bordeaux, Cadoret, p. 47.

48 Piriou. (1865) Considérations sur l'héméralopie et sur le scorbut. Extrait du rapport médical sur la station de la corvette à vapeur le Colbert au golfe du Mexique (1864–1865). Archives de Médecine Navale 4, 403–424; Vaucel (1891), p. 46; Guémar (1856).

49 Schwarz, E. (1861) Reise der Österreichischen Fregatte Novara um die Erde in den Jahren 1857, 1858, 1859 unter den Befehlen des Commodore B. von Wüllerstorf-Urbair. I. Medizinischer Theil. Wien, Kaiserlich-Königlichen Hof- und Staatdruckerei.

50 Hume, E. M., Krebs, H. A. (1949) Vitamin A requirements of human adults: an experimental study of vitamin A deprivation in man. A report of the Vitamin A Sub-Committee of the Accessory Food Factors Committee. Privy Council, Medical Research Council Special Report Series No. 264. London, His Majesty's Stationery Office, p. 139.

51 Schwarz (1861), p. 166.

52 Strambio, G. (1786) De pellagra. Observationes in Regio pellagrosorum nosocomio factæ a calendis junii anni MDCCLXXXIV usque ad finem anni MDCCLXXXV. Milan. (II, pp. 35, 40; III, pp. 37, 47, 49).

53 Billod, E. (1865) Traité de la pellagre d'après des observations recueillies en Italie et en France, suivi d'une enquête dans les asiles d'aliénés. Paris, Victor Masson et fils, pp. 167–168.

54 Roussel, T. (1866) Traité de la pellagre et des pseudo-pellagres. Paris, J. B. Ballières et fils, p. 27.

55 Simpson (1819), p. 66.

56 Fontan, J. (1882) De l'héméralopie tropicale. Recueil d'Ophtalmologie 4, 577–604.

57 Saltor, J. (1867) De la hemeralopia como efecto de los miasmas palúdicos, y de la afasia como consequencia de la caquexia palúdica en los niños. Compilador Médico 3, 357–359.

58 Junker, F. E. (1866) On night-blindness [letter]. Medical Times and Gazette (London) 2, 71; Graefe, A. (1859) Beiträge zum Wesen der Hemeralopie. Archiv für Ophthalmologie 5, 112–127.

59 Scarpa, A. (1818) A treatise on the principal diseases of the eyes. 2nd edition. Trans. J. Briggs. London, T. Cadell and W. Davies, p. 460; Alançon, D. M. (1835) Héméralopie sympathique observée chez un enfant de onze ans, et due à la présence d'entozoaires dans le tube intestinal. Journal des Connaissances Medico-Chirurgicales 3, 110–111.

60 Gayet. (1888) Héméralopie. In Dechambre, A. Dictionnaire encyclopédique des sciences médicales. 4 series, vol. 13. Paris, G. Masson, pp. 145–177.

61 Simpson (1819), p. 61.

62 Piriou (1865), pp. 410–411.

63 Boudet. (1873) De l'héméralopie et en particulier de l'héméralopie des pays chauds. Thèse pour le doctorat en médecine. Faculté de médecine de Montpellier. No. 9. Montpellier, Boehm et fils, p. 20.

64 Graefe (1859), p. 123.

65 Stellwag von Carion, K. (1868) Treatise on the diseases of the eye, including the anatomy of the organ. New York, William Wood, p. 656.

66 Gayet (1888), p. 160.

67 Fleury (1840), p. 54.

68 Eitner (1863), p. 248.

69 Jenkins, E. H. (1973) A history of the French navy from its beginnings to the present day. London, Macdonald and Jane's, p. 298.

70 Bonnafy, G. (1870) Considérations sur l'héméralopie. Thèse pour le doctorat en médecine. Faculté de médecine de Paris, no. 148. Paris, A. Parent, p. 28.

71 MacNamara, N. C. (1868). A manual of the diseases of the eye. London, John Churchill & Son, pp. 406–407.

72 Eitner (1863), p. 246.

73 Laveran (1858) Note sur la nature de l'héméralopie. Recueil de memoires de médecine, de chirurgie, et de Pharmacie Militaires 2 ser. 21:233–238; Gayet (1888), p. 155.

74 Hughes, R. (1987) The fatal shore. New York, Alfred A. Knopf, p. 96.

75 Trotter, T. (1799) Medicina nautica: an essay on the diseases of seamen. London, Longman and Rees, vol. 2., pp. 157–158.

76 Baugh, D. A. (1965) British naval administration in the age of Walpole. Princeton, Princeton University Press, pp. 375–376.

77 Lloyd & Coulter (1963), pp. 92–93.

78 Bryson (1859–1860), p. 40.

79 Le Roy de Méricourt (1867) Rapport sur les progrès de l'hygiène navale. Paris, L'Imprimerie Impériale, p. 24.

80 Fonssagrives, J. B. (1877) Traité d'hygiène navale. 2nd edition. Paris, J. B. Ballière et fils, p. 776.

81 Le Roy de Méricourt (1867), p. 29.

82 Fonssagrives (1877), pp. 894–895.

83 Gonzalez, P. M. (1805) Tratado de las enfermedades de la gente de mar, en que se exponen sus causas y los medios de precaverlas. Madrid, Imprenta Real, p. 13.

84 Clark, J. (1773) Observations on the diseases in long voyages to hot countries, and particularly on those which prevail in the East Indies. London, D. Wilson and E. Nicol, p. 332.

85 Bamfield (1819), p. 38.

86 Jayle, G. E., Ourgaud, A. G., Baisinger, L. F., Holmes, W. J. (1959) Night vision. Springfield, IL, Charles C. Thomas.

87 Admiral Digby's menu book, 1781 (Manuscript JOD/10).

Chapter 2

Paris in the Time of François Magendie

> There is a man who carries on his back newborn infants, in a padded box that can hold three of them. They are upright in their swaddling clothes, breathing the air from the top. The man stops only to eat and to let them suck a little milk. When he opens his box, he often finds one dead; he finishes his journey with the two others, impatient to be rid of his load. When he has left them at the hospital, he starts back at once, in order to resume the same job, by which he earns his daily bread [1].

So wrote Parisian playwright Louis-Sébastien Mercier in *Le Tableau de Paris* (Picture of Paris) in 1782. The man in *tableau* is a *meneur* – literally a ringleader and here a profiteer who gathers and sells abandoned babies. He transports his merchandise to the Hôpital des Enfants Trouvés in Paris. His buyers? Nuns, who pay him for the infants. Until new reforms of France's child welfare system were enacted in 1801, the Sisters of Charity paid *meneurs* for the orphaned infants delivered into their care. The more babies the sisters had in their custody, the greater the sums they could ask from the municipal administration that supported their charity [2]. Such were the costs of doing business.

The Paris of Mercier's *tableau* was about to burst: the French Revolution broke out in 1789, and calm would not return for a decade. And the dismal conditions many artists and writers portrayed would last through their own century and most of the next. Awash in dampness and mud, carriages and rag pickers, beggars and thieves, Paris was undergoing massive demographic changes. The urban population swelled with immigrants from the surrounding countryside. Between 1801 and 1851, the number of Parisians nearly doubled, from 547,756 to 1,053,261 (fig. 2.1) [3]. But the city's boundaries did not expand accordingly, and construction rates, too, failed to keep pace with the burgeoning populace. The result, of course, was a huge increase in density, but an uneven one. Of the city's twelve administrative sectors, the first arrondissement, which encompassed a posh part in the northwest of Paris, had a population of 15,383 people per square kilometer (fig. 2.2a); the Arcis quarter in the Seventh, in contrast, was home to some 93,781 people per square kilometer (fig. 2.2b) [4]. In the Seventh, where cheap boarding houses lined a warren of alleys, poor people lived ten to twenty crowded into one room. Conditions in the city's eastern arrondissements, including the twelfth where the Hôpital des Enfants Trouvés was located – on the rue d'Enfer (Hell Street) [5] – were no better.

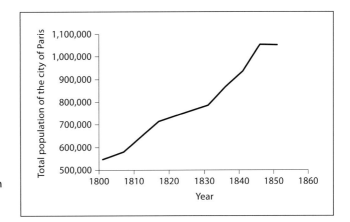

Fig. 2.1. Population of Paris in the first half of the nineteenth century [3].

Fig. 2.2. Comparison of the (**a**) 1st arrondissement with the (**b**) 7th arrondissement of Paris. Society for the Diffusion of Useful Knowledge. London, Chapman and Hall, 1834–35. Engraved by John Shury. Reproduced courtesy of the Geography and Map Division, Library of Congress.

Fig. 2.3. Abandonment of infants at the Hospice des Enfants Trouvés in Paris. The scene does not represent the typical scenario, as mothers usually came alone when they abandoned their infants. Lithograph by Pierre Émile Gigoux de Grandpré (b. 1826). Courtesy of Rachel Fuchs.

A portion of the city's exploding population consisted of pregnant young women who immigrated to Paris to turn their newborns over to charity. Accounts of unwanted babies born in Paris or deposited there from elsewhere go back centuries. In modern history, the rate of abandoned children in Paris multiplied many times over. By 1656, the city's general hospital had established a unit for these babies, and from 1670 through 1772, abandoned children admitted to what began as l'Hôpital des Enfants Trouvés (the Hospital for Found Children, i.e. foundlings) rose in number from 312 to 7,676 – a twenty-five-fold increase [6]. In the early nineteenth century, one infant in five in Paris was abandoned.

In the nineteenth century, most women who could not or would not keep their children were poor; many were unmarried, hence in disgrace. Their average age was twenty-six [7]. Those who had employment worked as domestics, seamstresses, or day laborers. Most of the babies they disposed of were a week old or younger.

A young woman could arrange to give her newborn over to l'Hôpital des Enfants Trouvés by having the delivery in the Maison d'Accouchement, a lying-in hospital. There, she would have to declare her intention. Once born, the infant would be transferred directly to the hospital, which, in 1801 was renamed l'Hospice des Enfants Trouvés. A wet nurse or priest might assist in the process. But if young mother-to-be went to the hospice alone, the Superior of the Sisters of Charity or a policeman might interrogate her and try to persuade her to change her mind. Or, as was more often the case, she could go at night after the delivery to the hospice's outside wall. There she would find a waist-high revolving drum – *un tour* – which, in one position, opened to the street, and, when turned, opened to the interior of the hospice (fig. 2.3). The *tour* allowed her to deposit her baby unseen. She might

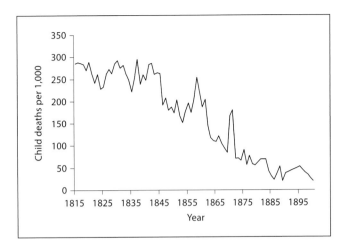

Fig. 2.4. Annual child mortality rate (deaths per 1,000 children admitted) at the Hospice des Enfants Trouvés (1815–1900) [9].

leave a note with her name and the baby's, and the date of the birth – or not. A bell cord beside the *tour* enabled her to alert the sisters within that a new baby had been left in their care.

Different but Hardly Better

After the Revolution, when the wealth of the Church and clergy in France was nationalized, the government charged that the nation, not the Church, must take responsibility for the moral education and physical well-being of abandoned children. Accordingly, it established procedures and regulations regarding the care of these orphans, and mandated the establishment of a foundling home in each *département* of France. L'Hospice des Enfants Trouvés in Paris became a part of a large governmental social welfare system that regulated and disbursed modest sums toward the care of France's orphans. Responsibility for tending the ever-growing number of abandoned children thus began a gradual shift toward government control. The shift became official and complete in 1811, when the French government decreed that the interior ministry, along with local departmental governments, had full responsibility for child welfare and abandoned children [8].

Physicians became important figures in the day-to-day operation of foundling homes, not only providing much-needed medical care, but also opening a path toward scientific breakthroughs in anatomy and nutrition (see chapter 3). Medical attention did not translate into a generation of healthy children or economically comfortable caretakers, however. During the first half of the nineteenth century, about 20–30% of the infants admitted to the Paris hospice died (fig. 2.4) [9] – an exceedingly high proportion, considering that most infants stayed in the hospice

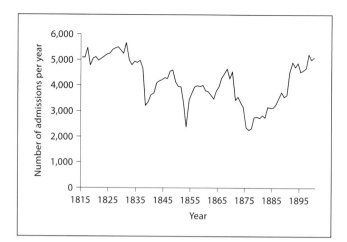

Fig. 2.5. Number of children admitted per year to the Hospice des Enfants Trouvés (1815–1900) [9].

for only a few days. Indeed, the infant mortality rate for all France at the time was high: about 20%. But the national rate, which encompassed a span of a full year – was dwarfed by the 20–30% mortality rate at the hospice, which covered just a few days.

As it continued to operate through the nineteenth century, the hospice served mainly as a way station and an agency for placing the abandoned infants in the care of rural wet nurses – at least, those babies deemed not to need medical attention. Rural settings were considered more salubrious than congested Paris. Removal to the country did not assure an infant's good health, however. The mortality rates for hospice infants under the care of rural wet nurses were two to three times greater than the infant mortality rates of France as a whole.

For the stronger babies' few days in the hospice before transfer, the hospice's fifteen to twenty-three full-time resident wet nurses provided nourishment until the rural wet nurses took charge [10]. And for the babies found needing medical care in the hospice infirmary, the hospice wet nurses were at hand. Both city and country wet nurses were paid for their services by the government, but poorly. Most wet nurses' pay was lower than that of ill-paid domestic servants or day laborers. In 1821, it started at 7 francs a month, with a 30-franc bonus for keeping an infant alive through her or his first birthday. Forty-five years later, a wet nurse's initial monthly pay had risen to a mere 12 francs [11]. For children who reached adolescence and beyond, a system of government-supported foster parents and apprenticeships was established (see textbox 2–1).

The number of children admitted to the Hospice des Enfants Trouvés fluctuated between about 2,200 and 5,700 during the 1800s (fig. 2.5). The rate of child abandonment in the eighteenth and nineteenth centuries rose and fell with the closely linked and highly volatile prices of wheat and bread [12].

Textbox 2–1. A precarious life for older orphans

Foster parents paid by the government usually took care of weaned abandoned babies and older abandoned orphans. Foster parents' monthly payments declined with the increasing age of the child or children in their care. The rationale for the decrease was that older children were deemed an economic asset, in that they could provide labor around the farm or elsewhere.

Before the expanded authority of l'Assistance Publique in 1852, abandoned children who reached the age of twelve were on their own if the foster parents did not want to keep them. Older abandoned children were therefore perceived to belong to the *classes dangereuses*; the fear was that these children would fall in with beggars, prostitutes, and criminals. After 1852, l'Assistance Publique took legal responsibility for abandoned children up to age twenty-one. Many of the older abandoned youths became agricultural day laborers, apprentices, or went into military service.

By law, all abandoned boys and other male orphans were required to serve in the military as a means of repaying the debt that they owed to the state for their care during infancy and childhood. Many of these youths were found unfit for the military because of health problems, including many they had acquired during infancy [13].

Bad Gets Worse

The poverty that plagued much of Paris during the first half of the nineteenth century seemed to become more deeply entrenched with the passage of time. One Parisian physician observed that the poor were subsisting on 'some crust of bread bought in the market, military biscuits sold by soldiers of the garrison, pork rind, or pork meat, horse meat [called 'beef of the poor'], cat meat, sometimes the entrails from the butcher, but more ordinarily dried beans and cheese' [14]. Another observer remarked that a common sight in Paris was 'the humiliating spectacle of unfortunates grubbing for dirty and unhealthy food in the refuse from the rich man's table in the garbage heaps on the streets, and even having to fight the animals for it' [15].

A major crisis hit – and by no means the last one – in 1815, when Mount Tambora in far-off Indonesia erupted. It was the largest and most consequential volcanic eruption then on record. Tambora expelled fifty cubic kilometers of dense rock, undercutting its peak from more than 4,300 m above sea level to 2,850 m (about 14,000 feet above sea level to less than 10,000 feet) [16]. The eruption ejected plumes to an altitude of forty-three kilometers and discharged sixty megatons of sulfur into the stratosphere. Ash blackened the sky and blanketed land and sea over an area of nearly 900,000 square kilometers (almost 35,000 square miles). The eruption's final death toll exceeded 71,000 – at first caused by lava flows, fires, and a tsunami, and later, by poor hygiene, diarrheal disease, and starvation.

The year 1816 came to be called 'the year without a summer'. Sulfur from the eruption formed an aerosol veil in the stratosphere, distorting normal weather patterns half a world away. The climactic changes brought massive crop failures and a subsistence crisis across Europe [17]. Famine ran rampant in 1816 and 1817. French villagers rioted, seizing grain before it was shipped out of their towns. They looted granaries and pillaged the houses of millers and civil officials [18]. From the German Rhineland came a report of 'wasted figures, scarcely resembling men, prowling around the fields in search of the unharvested and already half-rotten potatoes that never grew to maturity' [19]. In Switzerland's eastern cantons, beggars gathered, and the death rate rose steeply. In eastern Hungary and western Romania, 'People were eating grass, clover and maize-stalks, and perishing. . . of starvation and its attendant diseases'. Forty-four thousand people starved to death in that region [20].

In the rural region around Paris, peasants reacted with rage to the knowledge that much of what remained of a poor harvest was headed for the city. Some five thousand people laid siege to the town of Château-Thierry, where they emptied granaries and blocked barges on the Marne River. The insurrection spread to the surrounding region, where peasants armed themselves with sticks, bayonets, and sabers. The army deployed troops with orders to shoot – and shoot they did. Gunfire killed several of the rebels and wounded many more before the crowds dispersed [21].

First Steps in the Science of Public Health

The poverty of Paris added impetus to a growing interest in public health – a progressive line of inquiry that, like the Revolution, had roots in the Enlightenment of the eighteenth and early nineteenth centuries. Activists in the public health movement sought to improve health in the whole of society. Before the rise of public health as a discipline, and statistics as an essentially related field, the prevailing idea was that diseases and varying death rates resulted from temperature, winds, sunlight, humidity, elevation, and proximity to rivers and wetlands [22]. By investigating the influence of diet, behavior, occupations, climate, and 'all the nuances of social position', public health could 'exert a great influence on the development of the human spirit. . . enlighten the moralist and contribute to the noble task of diminishing the number of social ills' [23].

The physician Louis-René Villermé (1782–1863), a groundbreaking figure in the history of public health, conducted investigations that shined a harsh light on the health disparities between Paris's rich and poor (see textbox 2–2). The city made a living laboratory for Villermé, enabling him to study and compare inhabitants of the first arrondissement with those of the eighth, ninth, and twelfth – that is, the effects on people of living in opulent comfort versus living in squalor. Villermé found major contrasts within the ninth arrondissement itself. The annual death rate in the vermin-ridden rue de la Mortellerie was four and one-half times that of the commodious Ile Saint Louis on the other bank of the Seine [24]. Empathy and disapproval color descriptions of 'the poor, who have against them misery, excess work and want, and the rich, with their boredom, leisure, and excessively abundant diet' [25].

Textbox 2–2. Early observations of health disparities

The twelve arrondissements of Paris (fig. 2.6) served as a laboratory for Louis-René-Villermé, who reported the proportion of poor people living in each. He defined as 'poor' people who were living in untaxed, rented quarters. The arrondissement with the lowest proportion of poor people was the First, while the Eight, Ninth, and Twelfth had the highest proportion of poor people (fig. 2.7a) [26]. Villermé's statistical analyses showed a close association between higher levels of poverty and greater mortality (see fig. 2.7b). The mortality rates were 70% higher in the poor arrondissements than in the rich ones.

Villermé's contemporary, Louis-François Benoiston de Châteauneuf, approached the issue from another perspective. He demonstrated that the average lifespan of the rich was longer than that of the poor [27]. Diet and occupation were the causes he cited as underlying the difference. With longevity and mortality rates understood as the primary measures of a population's 'level of civilization', mortality studies became tools of nineteenth-century public health specialists 'engaging in the noble task of diminishing the number of social ills' [28].

While housing for Paris's growing population was largely neglected, city administrators tried to keep up with the increasing need for medical care. In 1817 nearly 41,000 patients were admitted to Paris hospitals [29]. The city's many hospitals and hospices were strongly centralized under the general council of hospitals after 1801 and under the Assistance Publique after 1849. The central authority maintained control over everything from admissions to pharmacies to food. Bread – 20,000 pounds of it a day – was supplied to hospitals by the Maison de Scipion, as were meat, wine, and candles. Some six thousand sheep, eighteen hundred head of adult cattle, and eight hundred calves were slaughtered each year to feed hospital patients [30].

Many of the poor were treated at the Hôtel Dieu, where the conditions before the Revolution were marked by a policy of putting as many beds as possible in one room and 'to put four, five, or six patients into one bed. . . convalescents together with the sick, the dying, and the dead. . .' [31]. After the Revolution, there were reforms such as enlargement of hospitals, provision of single beds, construction of new hospitals, and centralized administration. But mortality in the hospitals remained relatively high despite these initiatives. During the period 1815–1824, for example, 18% of the

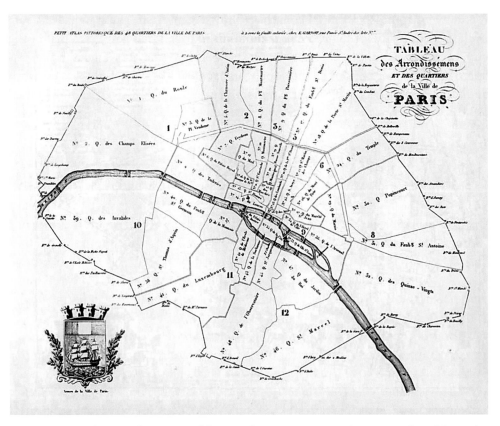

Fig. 2.6. The twelve arrondissements of the city of Paris in 1834. From Perrot, A. M. (1834) Petit atlas pittoresque des quarante-huit quartiers de la ville de Paris. Paris, Éditions de minuit [1960]. Reproduced courtesy of the Geography and Map Division, Library of Congress.

patients died who were admitted to the Hôtel Dieu, and 13% of patients died admitted to the Charité, which was regarded as the best hospital in Paris [32].

The statistical reports of Villermé captured the attention of writers such as Honoré de Balzac, who wrote of one arrondissement as a place in which '. . .two-thirds of the inhabitants lack firewood in winter, the district that sends most brats to the Foundling Hospital, the most sick to the Hôtel-Dieu, the most beggars onto the streets, the most ragpickers to the garbage dumps at the corner. . . the most indictments to the police courts' [33].

Poverty and hunger were thought to engender criminality, so the poor were widely viewed as the *les classes dangereuses* (the dangerous classes) [34]. Bread, regardless of nutritional content, filled bellies and thus deterred crime. Keeping the price of bread low, therefore, was regarded as key to keeping the peace; a high mortality rate was deemed to be a result of a high bread price [35]. A police prefect cautioned that, 'It is imperative to see that the price of bread does not go higher

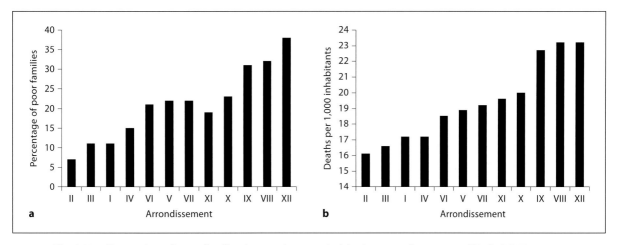

Fig. 2.7. a Proportion of poor families (untaxed properties) in the arrondissements of Paris (1817–1821). **b** At-home mortality in the arrondissements of Paris (1817–1821) [26].

than 3 sous the pound' [36]. In the 1820s, when Paris had some six hundred bakeries, bakers were not free to determine the price of their product: bread prices were a matter of the city's security, hence strictly regulated. The police and a four-member commission representing bakeries determined the current price of bread and issued a monthly edict [37].

From 1815 to 1816, the average monthly price in France for one hectoliter (2.7 bushels) of wheat nearly doubled. By 1817 the price was nearly triple that of two years before [38]. Bread prices, of course, reflected wheat costs. Thus, a four-pound loaf of bread came to exceed 20 sous – more than half again the police prefect's absolute maximum price. With no improvement in sight, the municipal government resorted to subsidizing a bread allowance for the poor: 227,399 people – nearly one-third of the city – qualified, including more than half the population of the twelfth arrondissement [39].

An increase in the price of bread was met with dread. Crowds would storm police stations and charity stations for bread tickets. Many people simply went hungry. In the infamous twelfth arrondissement, a police superintendent noted, 'I saw unfortunate fathers of families with their four or five children around them, without bread and with no notion how they were going to be able to support them next day' [40]. The cancellation of the government's bread allowance, plus the closing of soup kitchens only made matters worse. The ensuing riots garnered international publicity [41].

The bread riots reinforced the perceived link between poverty and crime, and the notion that the laboring classes and the poor were *les classes dangereuses*. Balzac sounded an alarm: 'The day when the mass of the unfortunate is greater than that of the wealthy, society will take on some totally different form. At this very moment England is threatened with a revolution of this kind. Taxes will become exorbitant for the poor. . . and on the day when 20 million men out of 30 million are dying of hunger, yellow leather breeches, cannons, and cavalry will be unable to do a thing about it' [42].

D'Arcet's Gelatin for the Needy and the Dietary Nitrogen Studies of Magendie

The poor of Paris had had their advocates since well before the Revolution and the eruption of Mount Tambora, and how to feed the poor efficiently was a longstanding concern of French philanthropists and intellectuals. In the early 1800s, scientists categorized foods as those with and without nitrogen. With its high nitrogen content, gelatin made mostly from bovine bones was proposed by some chemists and physicians as the economical solution to satisfying the nutritional needs of poor people in the streets and in hospitals. Gelatin therefore became a focus of food-related laboratory research. One physicist and inventor developed the 'steam digester' – a type of pressure cooker that could extract gelatin from bones. Another experimenter devised a way to extract gelatin by dissolving the calcium salts in bone with diluted acid, removing the excess acid, and boiling the extracted substance in water.

Antoine Cadet de Vaux (1743–1828), a chemist and advocate for the poor, declared that bone was 'a bouillon tablet formed by nature' and that 'one pound of bones can yield as much broth as six pounds of meat' [43]. The government issued a directive urging the greater use of gelatin to feed the sick and poor. Non-nutritional uses of bone incensed Vaux. 'One small box, one knife handle, or a dozen buttons made of bone', he railed, 'all of them correspond to a broth that has been robbed from the poor'.

Other scientists pushed Cadet de Vaux's interest in gelatin farther. One was the chemist Jean-Pierre-Joseph d'Arcet, who saw great practical and humanitarian, as well as nutritional, merit in Vaux's supposition. The prevailing notion in nutrition was that good health was dependent on nitrogenous foods, and he knew that the gelatin in beef bones and meat was rich in nitrogen – rich enough, he said, to 'render the food complete' even with relatively little protein [44]. He resolved, therefore, 'to consecrate all my life to add gelatin to the composition of the nitrogen-poor food which is distributed to the poor' [45].

In 1812, d'Arcet developed an inexpensive and efficient way to extract ample quantities of gelatin from bones and presented his technique to Paris's Société Philanthropique. He offered his process to prepare bouillon and soups for distribution to the poor and, notably, the sick in hospitals [46]. The philanthropic society, in turn, asked the faculty of medicine for a professional assessment of the nutritional value of d'Arcet's gelatin. A commission of distinguished scientists, including the noted chemist, Louis Vauquelin, was formed to study the matter.

The commission designed an experiment to be carried out with patients in the internal clinic of the Faculty of Medicine. The hospital staff was instructed to feed forty patients a diet that included bouillon prepared with d'Arcet's gelatin. They replaced three-fourths of the meat normally used in the preparation of beef bouillon

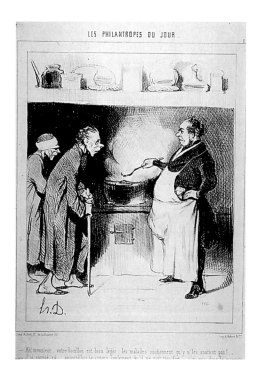

Fig. 2.8. Les philantropes du jour, by Honoré Daumier (1808–1879). From Le Charivari, September 15, 1844. '– Ah, sir, your broth is really weak... the patients complain that it does not sustain them! – I have corrected that... but today it might be a bit too strong... I added to the pot half a game of dominos more than yesterday... including the double six to give it a nice color...' During this era in Paris, dominos were made from bones. Reproduced courtesy of the National Library of Medicine.

with gelatin and carrots, onions, turnips, and other vegetables. The remaining three-fourths of the meat were served as a roast to the same patients. After three months on this diet, the patients were reported to be doing well. The commission therefore announced with certainty in 1814 that the gelatin was nourishing, easy to digest, healthful, and without any ill effects, and d'Arcet was hailed as having 'rendered a true service to humanity' [47].

Greatly encouraged, d'Arcet patented a gelatin-extraction process in 1817. In the apparatus he developed, beef bones were boiled in the morning for the preparation of normal broth; in the evening the same leftover bones were cleaned of cartilage, crushed, and put in cast iron cylinders, where they were exposed to steam for four days. During this time, water was trickled from the top over the broken bones and, at the bottom, diverted into containers. As the runoff cooled, it set into a gel that was then used a base to make broth. A handful of establishments, including the Hôtel Dieu, Hôpital Saint Louis, the Charité, Val de Grâce, and Hôtel de Monnaie, installed d'Arcet's steam-extraction equipment to prepare the broth to feed their indigent patients.

Life-sustaining though it may have been, d'Arcet's broth met with little enthusiasm from people to whom it was served. Indeed, some regarded it as a starvation diet, and it became a subject of suspicion and public ridicule (fig. 2.8). Balzac, for example, held it responsible for poor people's reluctance to enter hospitals, pointing out '. . .the popular conviction that hospitals starve their patients to death' [48]. Many critics saw no merits whatever in d'Arcet's brew: '. . .the broth takes from the gelatin all of its bad qualities. . . it spoils more easily than broth prepared the regular way. . . it is of a very disagreeable flavor which can inspire true disgust. . . it is less easily digested. . ... it contains smaller quantities of nutritious matter. . . [and] is of inferior quality than that contained in the ordinary broth'. Less scientific objections included that the broth made from d'Arcet's gelatin was simply, 'unappetizing, insipid, nauseating, and. . . detestable' [49]. Jean Nicolas Gannal, a chemist with considerable laboratory experience, was impressed by the fact that rats would never touch gelatin. He decided to test the edibility of gelatin by eating it himself and giving it to his family. But all the Gannals developed digestive problems, and they quit the experiment in disgust.

Worse than its lack of appeal, the gelatin broth was suspected of having insidious effects. At the Charité, some patients became gravely ill after being fed it, and one died [50]. Nor did it offer significant financial benefit. The Hôtel Dieu, for instance, saved only 7 francs 13 sous per day from its total hospital budget by serving its patients gelatin broth instead of regular broth [51]. The Hôtel Dieu was one of the first hospitals to give up manufacturing d'Arcet's gelatin for broth – which it did at the urging of its own physicians and pharmacists.

This turn of events dealt a blow not only to d'Arcet's purse but also to his professional pride and sense of ethics. He pointed out that a commission of the faculty of medicine had already certified the gelatin broth to be nutritious and healthful. For more than three years, d'Arcet argued, the Hôpital Saint Louis had fed his broth to twenty-nine thousand patients with great success [52]. But his opponents countered that patients at the Hôpital Saint Louis, which was a dermatological hospital, were not so gravely ill as those treated elsewhere. Indeed, in the original experiment of Vauquelin, the patients who were served roasted beef and vegetables along with some gelatin broth had fared generally well. So the question lingered: Wasn't the success in the Hôpital Saint Louis as likely attributable to the roast beef and vegetables as to any gelatin broth also served?

After hearing physicians' complaints, Alfred Donné, chief of the clinic at the Charité and one of the original proponents of gelatin, decided to conduct another, albeit small-scale, experiment. His subjects would be two dogs and himself. During six days on a regimen of gelatin and bread morning and evening, then his usual midday dinner, Donné was visited by a constant hunger. By the end, he had lost nearly one kilo. The dogs, meanwhile, simply refused the gelatin no matter how it was prepared, even on days when they were given nothing else [53].

Given that professional reputations as well as the well-being of thousands of people in need were at stake, the Academy of Sciences was eager to see the gelatin issue resolved. In 1831, the Academy reached a decision to study the problem in depth and appointed what came to be called the Gelatin Commission. D'Arcet was one of five scientists named to the commission, but, quite properly, he abstained from participating because he had a vested interest in its findings. Another prominent member of the commission was François Magendie (1783–1855), a noted physiologist and faculty member of the Collége de France. The rigorous study they developed, involving information from hospital patients and doctors, chemical analyses, testimonies, and tests on hundreds of animals, took a full ten years to complete (textbox 2–3).

At the time the Gelatin Commission was created, Magendie was already recognized as Paris's leading figure in experimental physiology. His renown grew out of a series of remarkable nutritional experiments begun in 1816 examining the importance of nitrogen in proper nutrition, and to this day, he is widely regarded as the first scientist to conduct long-term nutritional studies.

Textbox 2–3. The last word on d'Arcet's gelatin

The Gelatin Commission spent ten years studying the gelatin-nutrition question before publishing its findings in 1841. The report was sharply critical not only of d'Arcet's gelatin but also of the work by Cadet de Vaux that preceded it; the report excoriated Cadet de Vaux's observations as exaggerated and unscientific. As for the 1814 study commissioned by the Société Philanthropique, its findings were dismissed as motivated by unscientific ideals of fighting poverty and corrupted by unstated wishes to 'protect . . . innovators'. Furthermore, it completely disregarded the opinions of hospital patients made to eat the gelatin, who '. . .were unanimous in declaring their repugnance' [54]. The Commission also dismissed an earlier experiment by one Monsieur Robert, in which gelatin was fed to a dog. Robert's experiment apparently came to nothing, because 'the dog ran away'.

To determine the nutritional properties of gelatin scientifically, the Commission embarked on an ambitious set of experiments testing various diets on several hundred dogs (at the time, no standards gauging the nutritional value of particular foods had been established). The report noted that dogs would not touch gelatin in its pure form. When fed the gelatin broth as prepared in the Hôpital Saint Louis, the dogs might test the broth but would soon refuse it and begin to lose weight. When fed a broth prepared from meat by a Dutch company, however, the dogs gained weight. Dogs crossed over from one diet to the other showed a consistent effect: weight loss on the gelatin broth; weight gain on the meat-based broth.

The Commission then fed dogs extracts from bones; then tendons; then egg white mixed with bread; then fibrin extracted from beef blood; then butter; and finally, lard. The dogs would refuse to eat egg white; they would starve to death when fed fibrin – surprising findings, given that albumin in egg whites and fibrin in blood were considered nutritious. Then various mixtures of fibrin, gelatin, and albumin were tested, but the dogs eventually wasted away on all these.

One observation in the Commission report: 'As so often in research, unexpected results had contradicted every reasonable expectation'. The conclusion, drafted by François Magendie, was a rhetorical question: 'Have we not above all made it evident that science is still in its first steps in every aspect of the theory of nutrition?'

François Magendie was the product of a stern, doctrinaire upbringing that began in Bordeaux. His father, Antoine Magendie, was a surgeon, an ardent supporter of the political left, and an avid follower of the philosopher Jean-Jacques Rousseau. [55] Antoine raised his son according to tenets laid out in Rousseau's *Émile: or, On Education* (1762) [56]:

> The first education ought to be purely negative. It ought to consist not at all in teaching virtue or truth, but in shielding the heart from vice and the mind from error. If you could do nothing and allow nothing to be done; if you could bring your pupil sound and robust to the age of twelve years without his being able to distinguish his right hand from his left, from your very first lessons the eyes of his understanding would be open to reason; being without prejudice, without habit, he would have nothing in him which could counteract the effects of your endeavors. Soon he would become in your hands the wisest of men. You would have begun by doing nothing, and you would finally have produced a prodigy of education. . . leave your pupil to himself in perfect liberty, and observe what he does without saying anything to him [57].

When François was eight years old, his family moved to Paris, where his father gave himself over entirely to political activism, in time becoming mayor of the crowded tenth arrondissement, a member of the governing Commune, and, finally, administrator of public establishments. Antoine's tenure in the latter position ended badly: accused of negligent supervision of the *Maison des Enfants de la Patrie*, a foundling home, he was arrested [58]. Young François, meanwhile, was allowed to run free as the Rousseau doctrine would have it, with 'the eyes of his understanding' left open to reason and observation. He thus passed his tenth birthday with no schooling, unable to read or write, and accustomed to wandering barefooted through the streets of Paris. When he asked his father about wearing shoes, his father told him that if he wanted to wear shoes, he should make them.

The boy revolted, declaring that he wanted to be well shod and to go to school. His father finally relented and let François enter primary school, where at first he sat among much younger children. The young Magendie soon surpassed his classmates. At age fourteen, he entered the annual *concours* in which the best students of Paris competed and won first prize with his essay, 'On the Knowledge of the Rights of Man and of the Constitution' [59].

Magendie's rapid medical education was under way by the time the boy was sixteen, when he began working in Paris hospitals. At the Charité, he began to study with an old friend of his father's, Alexis Boyer, who had achieved fame as Napoleon's personal surgeon on the German campaign and was now Charité's second surgeon. Boyer recognized Magendie's talents and made him his prosector (i.e. dissector) for instruction in anatomy. Soon Magendie was lecturing in anatomy and giving his own lecture-demonstrations of the art of dissection. For most of his lectures, Magendie dissected human cadavers, but to teach comparative anatomy, he also dissected the bodies of goats and other animals.

While at the Hôpital Saint Louis, Magendie supported himself by teaching. He later recounted his frugal existence as an intern living in a Paris garret. After he paid

Fig. 2.9. François Magendie. Courtesy of the Archives of the Académie des sciences – Institut de France, Paris.

his necessary expenses, only five sous a day remained. He had a pet dog and recalled, 'So we shared. . . the dog was not fat, and neither was I' [60].

His fortunes changed abruptly one morning when a lawyer knocked at his door. Taken aback, the young doctor asked why a lawyer would call upon him, as he had no legal affairs and (probably recalling his father's experience) had not run afoul of the law. 'Nothing that can be disagreeable to you', replied the stranger. 'You have become heir to a sum of twenty thousand francs, and I am here to place it at your disposal.' Magendie's melancholy evaporated, and so, for the moment, did his austere existence. For the time being, he could live like a landed aristocrat and look the part (fig. 2.9). He purchased some fine horses and dogs, hired an elegant groom, and rented a livery stable near the Hôpital Saint Louis. He later recalled '. . .my whole recreation was literally centered on the stable.' Soon the twenty thousand francs were gone. He found it necessary to become a practicing physician 'in spite of himself' [61].

Magendie received his medical degree in 1808 and, in the following year, published his first paper, '*Quelques idées générales sur les phénomènes particuliers aux corps vivants*' (Some General Ideas on the Phenomena Peculiar to Living Bodies). In it, he stated that most '. . .physiological facts must be verified by new experiments, and this is the only means of bringing the physics [i.e. physical phenomena] of living bodies out of the state of imperfection in which it lies at present' [62].

The paper presented no new observations or findings; rather, it was a scientist's manifesto. As such, it was a posthumous challenge to the teachings of the late Xavier Bichat (1771–1802). An influential and respected teacher, Bichat had adhered to a doctrine of 'vital properties', in which the properties of tissues are related to their contractility, extensibility, and sensibility that followed the laws of physics. Physical forces outside the organism and the vital forces within the organisms were opposed to each other, thus, for Bichat, 'life is the ensemble of functions that resist death' [63]. Magendie had contempt for theories such as Bichat's and did not hesitate to put this sentiment into words. 'To express an opinion', he wrote, 'to believe, is nothing else than to be ignorant. . . one could with justice say to you "You believe, therefore you do not know"'. [64]. The path Magendie would follow was determined by a self-imposed necessity to establish scientific knowledge on the basis of direct observation.

In August of the 'year without a summer', Magendie presented a seminal experiment before his esteemed colleagues at the *Académie des Sciences*. He had set out to study the function of nitrogen in the diet by feeding dogs in his laboratory a regimen of sugar, gum Arabic, and olive oil or butter – substances with little or no nitrogen [65]. In explaining his objectives, he revealed his conviction that nutritional studies had become stuck in old ways – i.e. Bichat's – of investigating natural phenomena:

> It is not that nutrition has been neglected by physiologists; on the contrary, it has often been the object of conjectures and suppositions, sometimes very ingenious. But our knowledge is still too imperfect, so much that unless we follow experience step by step we cannot avoid getting lost. So all the hypotheses that have been made about nutrition are really nothing more than the expression of our current ignorance about the nature of this phenomenon, and they have solely the utility of more or less satisfying our imagination. It would be quite desirable, nevertheless, that we could obtain exact facts on the process of nutrition. This phenomenon is one of the most general and most important present in animals. Most diseases seem to be only alterations of it, consequently the discoveries that could be made on this domain would lead not only to advances in physiology, but even to useful applications in medicine – the goal towards which our labors must tend [66].

When Magendie fed a diet of sugar and distilled water to a dog, the animal grew thin, lost vigor and appetite, and, by the third week, developed ulcers on the cornea of each eye. The corneal ulcers eventually perforated, allowing the aqueous humor to escape: the result, although he did not record it, was almost certainly blindness. A daily allowance of three to four ounces of sugar was continued until the dog died at the end of one month. Magendie repeated the experiment several times, each time producing the same result.

He then conducted additional experiments in which he fed his dogs a regimen of gum arabic [67] and olive oil or butter. The outcome was the same for the dogs fed sugar – except that the corneal ulceration did not always occur [68]. The variability in the results may have been related to the nutritional state of the dogs before the experiments commenced.

What happened with some of Magendie's dogs was the same as the pathology seen in humans with severe vitamin A deficiency: he had induced what is known as xerophthalmia, or literally, a drying of the conjunctiva and cornea [69]. In humans, severe vitamin A deficiency leads to ulceration of the cornea, visual impairment, blindness, and in many sufferers, high mortality [70].

Among anti-vivisectionists and animal rights activists, Magendie may be remembered mostly for live anatomical dissections and his feeding experiments with dogs. In the scientific world, however, he is best known as a founder of experimental physiology and mentor to another noted pioneer in the field, Claude Bernard (1813–1878). In 1817, Magendie completed an important physiology textbook, *Précis élémentaire de physiologie* (Elementary Handbook of Physiology), in which he emphasized that experimental demonstration should replace theoretical discussions. Magendie's discoveries include the passive role of the stomach in vomiting, the digestive properties of pancreatic juice, the role of the liver in detoxification processes, demonstration of the hemodynamic importance of the large arteries, the role of the cerebellum in maintaining balance, the circulation of the cerebrospinal fluid from the brain cavities to the space around the spine, and the distinction between the motor and sensory roots of the spinal nerves [71].

The accomplishment of Magendie earned him a broad reputation among both colleagues of his own time and succeeding generations' learned men and women. Balzac created a fictitious character modeled on Magendie in the novel *La peau de chagrin* (1831). Balzac's Docteur Maugredie possessed 'a distinguished intellect, but [was] skeptical and contemptuous'. He 'claimed that the best medical system was to have none at all and to stick to the facts' [72].

References

1 Mercier, L. S. (1782) Le tableau de Paris. Nouvelle edition. Corrigée & augmentée. Amsterdam. Vol. 3, chap. 271, p. 239.

2 The national public treasury and, later, the local municipal authorities furnished support for the Hôpital des Enfants Trouvés. See Fuchs, R. G. (1984) Abandoned children: foundlings and child welfare in nineteenth-century France. Albany, State University of New York Press, p. 18.

3 Chevalier, L. (1973) Laboring classes and dangerous classes in Paris during the first half of the nineteenth century. New York, Howard Fertig, p. 183.

4 Demographia (2009) http://www.demographia.com/db-paris-arrondpre1860.htm [accessed 3 January 2009].

5 The rue d'Enfer is now the present-day Avenue Denfert Rouchereau.

6 Delasselle, C. (1978) Abandoned children in eighteenth-century Paris. In Forster, R., Ranum, O. (eds) Deviants and the abandoned in French society. Selections from the Annales economies, sociétés, civilizations. Volume IV. Baltimore, Johns Hopkins University Press, pp. 47–82.

7 Fuchs (1984), p. 87.

8 Fuchs (1984), p. 27.

9 Fuchs (1984), pp. 195–196.

10 Husson, A. (1862) Étude sur les hôpitaux, considérés sous le rapport de leur construction, de la distribution de leurs bâtiments, de l'ameublement, de l'hygiène et du service des salles des malades. Paris: Paul Dupont, Administration générale de l'Assistance publique à Paris, p. 311.

11 Fuchs (1984), p. 185.

12 Delasselle (1978), p. 71; Fuchs (1984), p. 98.

13 Fuchs (1984), pp. 260–261.

14 Leuret, [F.] (1836) Notice sur les indigens de la ville de Paris; suivie d'un rapport sur les améliorations dont est susceptible le service médical des bureaux de bienfaisance, fait au nom d'une commission. Annales d'hygiène publique et de médecine légale 15, 294–358.

15 Lescot, A. (1826) De la salubrité de la ville de Paris. Paris, Mme Huzard.

16 Stothers, R. B. (1984) The Great Tambora eruption in 1815 and its aftermath. Science 224, 1191–1198; Oppenheimer, C. (2003) Climatic, environmental and human consequences of the largest known historic eruption: Tambora volcano (Indonesia) 1815. Progress in Physical Geography 27, 230–259.

17 Post, J. D. (1977) The last great subsistence crisis in the western world. Baltimore, Johns Hopkins University Press.

18 Marjolin, R. (1933) Troubles provoqués en France pas la disette de 1816–1817. Revue d'histoire moderne 8, 423–460; Chabert, A. (1949) Essai sur les mouvements des prix et des revenus en France de 1798 à 1820. Paris, Librairie de Médicis, pp. 407–414.

19 von Clausewitz, C. (1922) Politische Schriften und Briefe. Edited by Hans Rothfels. Munich, Drei Masken, p. 190.

20 Macartney, C. A. (1969) The Habsburg Empire, 1790–1918. New York, Macmillan, p. 200.

21 Marjolin (1933), pp. 438–439.

22 La Berge, A. F. (1992) Mission and method: the early nineteenth-century French public health movement. Cambridge, Cambridge University Press.

23 Anon (1829) Prospectus. Annales d'hygiène publique et de médecine légale 1, vij, vj. Some of these precepts of public health were set forth by the founders of this new journal, which included Villermé.

24 Villermé, L. R. (1828) Mémoire sur la mortalité en France dans la classe aisée et dans la classe indigente. Mémoires de l'Académie royale de médecine 1, 51–98. The rue de la Mortellerie was located at the rue de l'Hotel-de-Ville in the present-day 4th arrondissement. The disparity in death rates among those dying at home between the two streets was greatly underreported, since a single room in the lodgings on the rue de la Mortellerie might have had nearly two dozen inhabitants. Moreover, the poor did not appear on the bills of mortality, since they were usually taken to hospitals to die. The wealthy, in contrast, would most likely die at home, among family members and servants, and under a physician's care.

25 Anon (1824) France. Paris, 9 décembre. Journal des débats politiques et litteraires. December 10, 1824.

26 Villermé, L. R. (1830) De la mortalité dans les divers quartiers de la ville de Paris, et des causes qui la rendent très différente dans plusieurs d'entre eux, ainsi que dans les divers quartiers de beaucoup de grandes villes. Annales d'hygiène publique et de médecine légale 3, 294–341; data from p. 310, corrected for typographical errors.

27 Benoiston de Chateauneuf, [L. F.] (1830) De la durée de la vie chez le riche et chez le pauvre. Annales d'hygiène publique et de médecine légale 3, 5–15.

28 d'Ivernois, F. (1834) Sur la mortalité proportionnelle de quelques populations, considérée comme mesure de leur aisance et de leur civilisation. Annales d'hygiène publique et de médecine légale 12, 231–200; d'Ivernois, F. (1834) Première et seconde lettres adressées à M. Villermé, sur la mortalité proportionnelle des peuples, considérées comme mesure de leur aisance et de leur civilisation. Annales d'hygiène publique et de médecine légale 12, 200–203.

29 Bouchardat, A. (1853) Nouveau formulaire magistral, précédé d'une notice sur les hôpitaux de Paris. Paris, Germer Baillière, p. 18.

30 Weiner, D. B. (1993) The citizen-patient in revolutionary and imperial Paris. Baltimore, Johns Hopkins University Press, p. 60.

31 Ackerknecht, E. H. (1967) Medicine at the Paris hospital 1794–1848. Baltimore, Johns Hopkins University Press, p. 19.

32 Meding, H. L. (1853) Manuel du Paris médical: recueil des renseignements historiques, statistiques, administratifs et scientifiques sur les hôpiteaux et hospices civils et militaires. Paris, J. B. Ballière, p. 61; Ackerknecht (1967), p. 20.

33 Balzac, H. (1856) Scènes de la vie privée. Honorine. Le colonel Chabert. La messe de l'athée. L'interdiction. Pierre Grassou. Paris, Librairie Nouvelle, p. 207.

34 Chevalier (1973), p. 120; Frégier, M. A. (1840) Des classes dangereuses de la population dans les grandes villes, et des moyens de les rendre meilleures. 2 vols. Paris, J. B. Baillière.

35 Mêlier. (1843) Des subsistances envisagées dans leurs rapports avec les maladies et la mortalité. Annales d'hygiène publique et de médecine légale 29, 305–331.

36 Chevalier (1973), p. 262.

37 Tulard, J. (1964) La Préfecture de police sous la Monarchie de Juillet suivi d'un inventaire sommaire et d'extraits des rapports de la Préfecture de police conservés aux archives nationales. Paris, Imprimerie Municipale, p. 97.

38 Labrousse, E., Romano, R., Dreyfus, F. G. (1970) Le prix du froment en France au temps de la monnaie stable (1726–1913). Paris, S.E.V.P.E.N., diagram 2.

39 Anon (1830) Budget de la ville de Paris pour 1830. Journal des débats politiques et littéraires November 2, 1830, p. 2. The need for such a large bread allowance occurred in 1830 with a spike in the price of bread.

40 Chevalier (1973), p. 266–267.

41 Anon (1846) Bread riots in Paris. The Illustrated London News 9, 226.

42 Balzac, H. (1995). Le code des gens honnêtes, ou L'art de ne pas être dupe des fripons. Paris, L'École des loisirs, p. 18.

43 Cadet de Vaux, A. A. (1803) Mémoire sur la gélatine des os, et son application à l'économie alimentaire, privée et publique, et principalement à l'économie de l'homme malade et indigent. Paris, Marchant.

44 Holmes, F. L. (1973) Claude Bernard and animal chemistry: the emergence of a scientist. Cambridge, Massachusetts, Harvard University Press, p. 8, citing Gelatin Dossier, D'Arcet, 'Note', No. 61, AAdSc.

45 Coulier. (1881) Gélatine. In Dechambre (ed.). Dictionnaire encyclopédique des sciences médicales. 4th series, vol. 7, Paris, G. Masson, P. Asselin, pp. 215–233.

46 d'Arcet, [J. P. J.] (1814) Rapport sur un travail ayant pour objet l'extraction de la gélatine des os, et son application aux différens usages économiques. Journal de médecine, chirurgie et pharmacie 31, 352–359.

47 Leroux, [J. J.], Dubois, [P.], Pelletan, Dumeril, [A. M. C.], Vauquelin, [L. N.] (1814) Rapports sur un travail de M. Darcet, ayant pour object l'extraction de la gelatine des os, et son application aux différens usages économiques. Journal de médecine, chirurgie et pharmacie 31, 352–359.

48 Balzac, H. (1856) Le cousin Pons. Paris, Librairie Nouvelle, p. 233

49 Coulier (1881), p. 223.

50 Anon (1833) Sur l'emploi de la gélatine comme substance alimentaire dans les hôpitaux et chez les indigens. Gazette médicale de Paris 2nd series, 1, 101–102.

51 Coulier (1881), p. 223.

52 D'Arcet, [J. P. J.] (1833) Réponse de M. D'Arcet a l'article relatif à l'emploi alimentaire de la gélatine, inséré dans le numéro 16 de la Gazette Médicale. Gazette médicale de Paris 2nd series, 1, 156.

53 Donne, A. (1835) Mémoire sur l'emploi de la gélatine comme substance alimentaire. Paris, Rignoux.

54 Magendie, F. (1841) Rapport fait à l'Académie des Sciences au nom de la Commission dite de la gélatine. Compte-rendus des séances de l'Académie des Sciences 13, 237–283.

55 Genty, M. (1935) Magendie (François) (1783–1855). In Lereboullet, P. (ed.) Les biographies médicales: notes pour server à l'histoire de la médecine et des grands médecins. Paris, J. B. Balliere et fils, pp. 113–144.

56 Flourens, P. (1858) Éloge historique de François Magendie. Paris, Garnier Frères, p. 3.

57 Rousseau, J. J. (1762) Émile: ou, de l'éducation. Leipzig, M. G. Weidmann & Reich, pp. 192–193.

58 Genty (1935), p. 113.

59 Flourens (1858), p. 5.

60 Flourens (1858), p. 8.

61 Flourens (1858), pp. 10–11.

62 Magendie, [F.] (1809) Quelques idées générales sur les phénomènes particuliers aux corps vivans. Bulletin des sciences médicales de la societe médicale d'émulation 4, 145–170.

63 Bichat, X. (1800) Recherches physiologiques sur la vie et la mort. Paris. Chez Brosson, Gabon et Cie. Bichat held that there were two fundamental properties, sensibility and contractility, and the list of 'vital properties' were (1) animal sensibility, (2) organic sensibility, (3) animal contractility, (4) sensible organic contractility, and (5) insensible organic contractility.

64 Dawson, P. M. (1908) A biography of François Magendie. Brooklyn, New York, Albert T. Huntington, p. 17.

65 Magendie, F. (1816a) Mémoire sur les propriétés nutritives des substances qui ne contiennent pas d'azote. Bulletin des sciences par la Société Philomatique de Paris 4, 137–138; Carpenter, K. J. (1994) Protein and energy: a study of changing ideas in nutrition. Cambridge, Cambridge University Press, pp. 27–30.

66 Magendie, F. (1816b) Mémoire sur les propriétés nutritives des substances qui ne contiennent pas d'azote. Annales de chimie et de physique (sér. 2) 3, 66–77.

67 A natural, edible gum made from the hardened sap from Acacia trees and used as an adhesive, in foods, polishes, inks, and medicines.

68 Magendie (1816a), p. 138.

69 McCay C. M. (1930) Was Magendie the first student of vitamins? Science 71, 315.

70 Semba, R. D. (2007) Handbook of nutrition and ophthalmology. Totowa, New Jersey, Humana Press, pp. 27–30. The development of corneal ulceration in dogs fed butter is a puzzle, since butter is generally a rich form of vitamin A. Olmsted conjectured that the source of the butter may have been cows that were kept in a stable and fed no fresh grass, 'in which case their butter would have a low vitamin A content' in Olmsted, J. M. D. (1944) François Magendie: pioneer in experimental physiology and scientific medicine in XIX century France. New York, Schuman's, p. 68. This may well have been the case, given the conditions of the spring and summer across France in 1816.

71 Grmek, M. D. François Magendie. In Gillispie, C.C. (ed.) Dictionary of Scientific Biography. Volume IX. New York Charles Scribner's Sons, 1974, pp. 6–11.

72 Balzac, H. (1838) La peau de chagrin. Paris, H. Delloye, Victor Lecou, p. 340.

Chapter 3

Deprivation Provides a Laboratory

I have seen many infants who were reduced to complete marasmus by gastro-intestinal disorders of long duration, affected. . . with softening of the cornea. . . followed by a perforation and a discharge of the humors of the eye and the crystalline lens. This type of spontaneous softening reminds me of the fact noticed by M. Magendie in a dog, which, being fed for a long time with sugar, died after having been reduced to great emaciation.

From Charles-Michel Billard, *Traité des maladies des enfants nouveax-nés et À la mamelle, 1828* [1].

During the decades of reform after the French Revolution, Paris's Hospice des Enfants Trouvés served as a teaching hospital. The Sisters of Charity tended Paris's proliferation of abandoned babies left in their charge under the directorship of a distinguished pediatrician, Jacques-François Baron (1782–1849). Young physicians-in-training, notably Charles-Michel Billard (1800–1832), accompanied Baron on daily clinical rounds in the nursery (textbox 3–1).

Perhaps Billard cannot have been surprised to find the care of the hospice's abandoned infants to be methodical and attentive but not very successful in starting them on the road to healthy lives. On any given day, the hospice held on average about three hundred infants, but it had only one hundred seventy cribs – eighty cribs in the nursery and ninety in the adjacent infirmaries. The nursery was a former chapel with a high, vaulted ceiling – a cavernous space that was virtually impossible to heat despite a large fireplace and two stoves added at the physicians' insistence. Babies were doubled up in many of the nursery cribs, which were arranged in four rows separated by narrow aisles [2].

Before the doctors' rounds, the sisters bathed and dressed the infants. With the new arrivals, once they were clean and clothed the sisters took them directly to the chapel to be baptized: immediate baptism was essential, as death might well be imminent. Only after these preliminaries were the infants ready for the scrutiny of Baron, Billard, and the other young doctors under the physician-director's tutelage.

During rounds, Baron and his students passed pass from crib to crib, noting the babies' skin color, taking their temperatures, and assessing the firmness of their skin. On the basis of their observations, the physicians decided which infants were healthy enough to stay in the nursery for the time being, and which must be transferred to one of the hospice infirmaries for special treatment. Severely wasted babies suffering from diarrhea were common. The doctors paid particular attention to the infants' eyes, many of which showed small, well-defined ulcers in the cornea; when the ulcers perforated, fluid (aqueous humor) escaped. Otherwise, the eyes were clear and showed no discharge or infection [3].

Working under Baron, Billard spent the year 1826 as an *élève interne* [4] at the hospice.

He took special note of the weights of the newborns in the hospice. While low, birth weights there were not markedly different from the national average at the time: 2,268 to 2,495 g (about 5–5.5 pounds). (By present-day standards, however, the average nineteenth-century French birth weight is distinctly low for full-term births: the World Health Organization today defines low birth weight as less than 2,500 g [5.5 pounds]) [5]. Of the 5,392 infants admitted to the hospice that year, 1,414 died – a mortality rate of 26.2% [6]. This sad fact enabled a great many autopsies to be conducted and made the hospice a fertile place in which to investigate the pathology of diseases that can afflict newborns and infants.

The observations Billard was able to make over that year provided him with basis for a landmark pediatric textbook. His *Traité des maladies des enfans nouveax-nés et à la mamelle* (Treatise on the Diseases of Newborn and Nursing Infants) was published in 1828. Just one innovative feature of this work, which would become standard in comparable works to follow, was that its contents were organized by pathological anatomy rather than by clinical symptoms. In his treatise, Billard provided an important clinical description of the early stages of the corneal ulceration and keratomalacia (also called melting of the cornea) that would later become known to occur in infants with vitamin A deficiency. He was quick to see a parallel with the corneal ulceration that François Magendie had induced a decade before in dogs fed an experimental diet of sugar and water (see Chapter 2). And he posed the critical question of whether the infants' corneal softening could be attributed to a nutritional deficiency:

I have seen many infants who were reduced to complete marasmus [emaciation] by gastrointestinal disorders of long duration, affected, without palpebral inflammation (i.e. between upper and lower eyelids), with softening of the cornea, which was followed by a perforation and a discharge of the humors of the eye and the crystalline lens. This type of spontaneous softening reminds me of the fact noticed by M. Magendie in a dog, which, being fed for a long time with sugar, died after having been reduced to great emaciation. 'There appeared', says M. Magendie, 'on one eye, and afterwards on the other, a small ulcer on the center of the transparent cornea; it rapidly increased in size, and at the end of a few days it was about a *ligne* (2.25 mm) in diameter, and its depth increased in the same proportion; the cornea was soon perforated, and the humors of the eye escaped. This singular phenomenon was accompanied by with an abundant secretion from the glands of the eyelids.' Was a defect in alimentation a cause of the softening of the cornea? [7].

Textbox 3–1. C-M Billard, zealot with exceptional insights

Born in the Loire region near the town of Angers, Charles-Michel Billard and his brother were orphaned quite young and raised by an aunt [8]. Charles-Michel entered medical school in Angers at age nineteen. Shortly after he began medical studies, he became an *élève externe* at the Hôtel Dieu, the teaching hospital of the medical school. By 1823, he had already had considerable experience in conducting autopsies, and the knowledge he gained this way served him well when he entered an essay on the gastrointestinal mucosa in a Paris Athenaeum of Medicine competition. His essay won first prize, which came with a handsome 300 francs. That sum, plus a modest inheritance, enabled young Billard to pursue medical studies in Paris, where his successes continued. As a student, he augmented his means by translating texts into French, including *A Treatise on the Diseases of the Eye* by the eminent British ophthalmologist, Sir William Lawrence – a task that provided useful guidance for Billard's efforts to understand the ailments of sick infants.

After a series of top scores in competitions and an externship at the Salpêtrière, Billard found himself free to choose where he wished to spend the next phase of his training. His choice was to be an *élève interne* at the Hospice des Enfants Trouvés. The job at the hospice carried little status, but Billard had been inspired by anatomical pathologist Giovanni Morgagni, who, many decades before, had lamented, 'How vast and new is the space that is still open before us for the study of the diseases of young children!' [9]. In working with the abandoned children at the hospice, Billard could observe children closely, try to ascertain the causes of diseases, and in some of the many instances when no cure could be found, conduct autopsies.

In his pediatric textbook, Billard wrote, 'In this manner has the wish of Morgagni been fulfilled' [10]. But Billard's could not fulfill his own wish. His career and his service to the outcasts and downtrodden of Paris and then, briefly, of Angers to which he returned, were cut short when he died at age thirty-one of tuberculosis. He had followed a vital thread in François Magendie's line of inquiry. Other scientist-physicians would pick up the thread that slipped from Billard's grasp.

'A Defect in Alimentation...'

Of the numerous hospice babies who died, many did so of infectious diseases such as diarrhea or measles (see Chapter 7) – conditions and outcomes that would eventually become recognized as characteristic of severe clinical vitamin A deficiency. Given the infant feeding regimen the hospice was compelled to follow, much of the high mortality in the hospice quite likely resulted from low vitamin A intake. Inevitably, the quantity and nutritional value of wet nurses' breast milk come into question.

Simply put, supply could not come close to meeting demand. Of the hospice's dozen or so resident wet nurses, some were nursing their own infants as well as the hospice's several hundred foundlings. While one wet nurse could produce roughly 1.4 liters (three pints) of milk a day, each of the four to seven infants reliant on her for their nutrition got no more than 0.23 to 0.35 liters per day – a mere 30 to 40 percent of the milk the babies needed. The infants were literally being starved of vitamin A, as they would receive less than one-quarter of the vitamin A consumed per day by healthy infants [11]. The inadequate intake of vitamin A, combined with diarrhea caused by infectious viruses or bacteria, became a deadly combination.

Confronting a nursery full of hungry babies, the wet nurses supplemented breast milk feedings with either cow or goat milk or a mixture of bread and milk diluted with water. The milk was stored at room temperature in large jars in the center of the room, where it was exposed to dust raised by dry sweeping several times a day [12]. In addition, the cow or goat milk was raw, as sterilization with heat to curb bacterial contamination (i.e. pasteurization) was still unknown [13]. Nor were the benefits of refrigeration even if it were possible.

The low birth weights observed in France during the early 1800s indicate overall poor maternal nutrition during pregnancy. Being generally poor, the French women who abandoned their infants presumably lived on inadequate diets both before and during pregnancy. The mothers' meager, vitamin A-poor diets would have led to vitamin A-deprived fetuses in utero and vitamin A-deficient newborns.

Before an infant is weaned, she has two main sources of vitamin A. The first is her liver, in which a store of vitamin A has built up during gestation; vitamin A is transferred across the placenta from the mother's to the fetus's circulation, which, in turn, carries it to the fetal liver. The second source of vitamin A for an infant is the milk she takes in, whether from her mother's breast, a wet nurse's, or a bottle. For a breast-fed newborn, the store of vitamin A in her liver is a function of her mother's or wet nurse's intake of vitamin A-rich foods both during and after pregnancy. After the birth, the vitamin A content of breast milk is a reflection of the vitamin A status of the woman providing the milk. Because a pregnant woman requires more vitamin A than does a woman who is not pregnant, a mother-to-be is likelier to be deficient in vitamin A. One manifestation of vitamin A deficiency that was common in pregnant women is night blindness; this condition (see Chapter 1) was noted not only in rural France but also elsewhere in Western Europe and Russia [14]. Whether night blindness disproportionately affected French women who abandoned their babies is not known.

By all accounts, the diet of the resident wet nurses in the Hospice des Enfants Trouvés was poor in vitamin A. The hospice was required to provide the wet nurses with a certain amount of bread, wine, and meat. When a special commission was established to investigate the high mortality rates of hospice infants, it recommended that wet nurses' bread ration be augmented and that the number of resident wet nurses be increased from the usual twelve to twenty-three to forty.

After 1859, as part of what was heralded as a great nutritional improvement for resident wet nurses, their ration of wine was increased from 220 to 320 milliliters (7.4 to 10.8 fluid ounces) per day. In addition to their regular ration of boiled meat, they were given an extra 200 grams (seven ounces) of roast meat per day [15]. The new reforms did not change the amount of vitamin A in the diet, however, nor did it bring improved health prospects for hospice infants. The persistently high death rates among the infants can be attributed in a large part to poor hygiene, to the inadequacy of their nutritional intake, and specifically, to a lack of adequate vitamin A [16]. According to Victor Hutinel, physician-director of the hospice in the latter part of the nineteenth century:

When the infant cried, the nuns prepared something for her to suck upon made of a pinch of biscuit crumbs wrapped in a cloth rag poultice moistened by a sticky syrup exposed to all germs. Soon a rash appeared, a stubbornly virulent rash, then vomiting and diarrhea. Weight loss was hundreds of grams per day, and the drama of *athrepsie* [17] or dehydration followed, slowly or rapidly, according to the season. In summer, an infant was transformed in twelve hours. In the morning, she was pink and lively; at night, she was bluish, cold, and moribund; she had 'turned' as the nuns said, who likened this development to fermentation [18].

When l'Assistance Publique took responsibility for older abandoned children in orphanages, the children remained under the care of hospice physicians. At the Hospice des Enfants Assistés in Bordeaux, Pierre Bitot (1822–1888) looked after the health of growing orphans. Like Baron, Billard, and their colleagues, Bitot paid particular attention to the children's eyes.

Born in a small town southeast of Bordeaux, Bitot did his medical studies in that city before going into practice in Paris in 1848 [19]. He returned to Bordeaux in 1849 to accept the position of physician for the Bureaux de Charité and soon became professor of anatomy at the medical school where he had studied. Two years later, he became chief surgeon of Bordeaux orphanages, which post enabled him to resume his clinical observations, notably of xerophthalmia, i.e. dryness of the cornea and conjunctiva.

Bitot was particularly struck by the frequency of night blindness accompanied by the appearance of distinctive silvery white or pearly patches on the affected children's conjunctivas. By 1861, he had observed, in a total population of four hundred children, twenty-nine with both night blindness and the pearly patches on the eyes (fig. 3.1). Nineteen were boys ages six to sixteen; ten were girls between ten and nineteen [20]. (Large epidemiological surveys conducted in the 1970s and 1980s have consistently shown that boys are at higher risk of vitamin A deficiency than girls. Night blindness tends to be more common in preschool-aged children, as the diet is usually better among school-aged children. The presence of night blindness in older children and teenagers is strong indicator of the lack of vitamin A in the diet of the hospice.)

Bitot's observations coincided almost exactly with those of a Portuguese ophthalmologist, João Clemente Mendes, who described night blindness and xerophthalmia in a Lisbon orphanage [21]. And earlier clinical descriptions had been made of pathologic changes of the conjunctiva associated with night blindness [22]. But none of the descriptions are complete as those Bitot's. The observations and connections that Bitot made, and the follow-ups he conducted, yielded important results.

Bitot produced the first detailed description of where the conjunctival patches, or lesions, appear and the clinical course they follow. If scraped off, they reappear in the same places. Bitot examined the lesions microscopically and found 'a yet undescribed alteration, a specific squamous production of the conjunctival epithelium'. In another phase of his investigations, he examined children who had left the hospice. These children were affected with night blindness and the eye lesions while they lived in the hospice,

Fig. 3.1. Bitot's spots. The foamy raised lesion on the temporal conjunctiva, or Bitot's spot, is considered diagnostic for vitamin A deficiency. Courtesy of Alfred Sommer.

but once they were living outside the hospice, their symptoms and lesions resolved. (The eye lesions likely resolved in the children who left the hospice because they received a better diet with more vitamin A than the unvaried diet they received in the hospice. Bitot did not, however, make a connection between diet and the eye lesions.)

Pearly patches on the conjunctiva – today called 'Bitot's spots' – are now considered pathognomonic, i.e. diagnostic, for clinical vitamin A deficiency.

Gains in Nutrition, Then a Disastrous Reversal

Starting in the mid-nineteenth century, French mortality rates both nationwide and in the Paris Hospice des Enfants Trouvés slowly declined (fig. 2.4). The gradual improvement cannot be attributed to better hygiene or nursing practices in the hospice. These did not change greatly, nor did the treatment of sick infants. But a slow, general improvement in the overall nutritional status of the French population may account for the decline in mortality. Among pregnant women, a gradual improvement in diet would be associated with lower mortality rate of hospice infants. Stunted growth, an index of chronic malnutrition, also decreased steadily; in particular, the proportion of stunted men conscripted into the army diminished [23].

But a sharp peak in deaths abruptly interrupted the general decrease in mortality among abandoned children in the hospice. The dates of the peak, 1870–1871, correspond directly to those of a period of extraordinary suffering among the people of Paris, especially the poor. The Franco-Prussian War and the Prussians' Siege of Paris brought ills of monumental proportion.

In July 1870, with diplomatic relations with Prussia already under great strain, France's Emperor Napoleon III declared war. King Wilhelm II, the reticent monarch of a newly unified Germany led by Prussia, might have preferred to resist the bait. But Otto von Bismarck, the king's chancellor and literally the power behind the German throne, relished the challenge to assert the new nation's imperial power. Thanks to the attention Bismarck had lavished on arms and the army, an efficient German offensive crushed France's poorly organized and ill-equipped troops in August and, on

September 1, trapped the French emperor and field army in cities in France's northeast. Napoleon III and 83,000 French troops surrendered on September 2, and the French Empire was declared fallen.

A so-called Government of National Defense assumed power in Paris on September 4. It attempted to negotiate peace treaty favorable to France, but Bismarck demanded that regions of Lorraine and Alsace on the west bank of the Rhine River be ceded to Germany. The new French government rebuffed this demand out of hand. In response, Bismarck decided not to bombard the French capital but, rather, to starve the city by laying siege.

Cut off from the world, Paris presumably would surrender in weeks. The German chancellor snidely quipped that the bourgeoisie of Paris would break after 'eight days without *café au lait*' [24].

As the Prussian army approached Paris, livestock in the vicinity were quickly brought into the city. The poor were allowed to pick potatoes and other vegetables to keep them from the enemy. By the end of September, it was noted with some satisfaction that there were 24,000 cattle, 150,000 sheep, and 6,000 hogs within Paris [25]. The government considered this supply sufficient for Paris's estimated population of 1.5 million. The Parisians, thinking they were well stocked with food, did not modify their eating habits.

Paris held its ground, and Bismarck and his army held theirs. But as the siege wore on, supplies of certain foods began to become short. The French government soon realized it had made a major miscalculation. A census conducted in late December found that there were actually more than two million people living in Paris not counting the armed forces [26].

The first foods to become scarce were butter, milk, cheese, and eggs, and fresh vegetables and fruits. As prices for these foods rose, the poor were the first to feel the effects; the average French male laborer was making less than five francs a day, and the poorest were living on less than one franc a day (table 3.1) [27]. Beef and pork quickly disappeared from the market, so horsemeat became the main source of protein: seventy thousand horses were butchered for eating during the siege. Eventually, horsemeat ran low. Butchers began to deal in dog and cat meat, and later sold rats caught in the alleys and sewers.

At the local political clubs, which usually met in the evenings for discussion, lack of food was a common topic. In early December, the French writer and critic, Edmond de Goncourt, commented, 'Hunger begins and famine is on the horizon' [28]. With starvation looming, acts of desperation in the streets became increasingly common. An American journalist serving in the Parisian ambulance service described one:

At 3 o'clock in the afternoon, December 27, a half-starved horse fell in one of the streets of the Faubourg La Villette, and the driver, after calling to his assistance a few city loungers, was unable to raise him upon his feet. A butcher came and proposed to purchase him for the municipal slaughter house. Immediately [the horse] received the *coup de grace* and the butcher went in search of a cart. He had scarcely turned his back when the crowd raised a *hurrah* and precipitated themselves upon the horse. Every one wished a piece of him. Men, women and children gathered around the carcass

Table 3.1. Food prices during the siege of Paris, 1870

Food	Price in francs*	
	September 19–30, 1870 first two weeks of siege	December 10–24, 1870
Butter	4.00 per pound	35.00 per pound
Eggs	1.80 per dozen	24.00 per dozen
Cheese	2.00 per pound	30.00 per pound
Fowl	6.00 each	26.00 each
Rabbit	8.00 each	40.00 each
Pork, fresh	1.10 per pound	unobtainable
Potatoes	2.75 per bushel	15.00 per bushel
Cat	not consumed	6.00 per pound

* The average wage of a worker in Paris was less than five francs per day in 1870.

like a group of savages with knives in their hands. In less than twenty minutes there remained nothing of the animal but the hide and bones [29].

The situation only worsened as the months passed. Few of the animals in the Paris zoo were spared during the siege: butchers purchased exotic birds, antelopes, camels, yaks, zebras, bears, wolves, and kangaroos to sell in their *boucheries*. Even the beloved elephants, Castor and Pollux, were shot and slaughtered for their meat.

For the rich of Paris, the siege was but a minor inconvenience: for the most part, they had well-stocked pantries and could afford exorbitant food prices. It was noted, 'In the expensive cafes of the Boulevards, feasts worthy of Lucullus are still served' [30]. With the zoos rather than the farm now providing meat for the table, the journalist remarked, 'The epicureans of Paris have now an opportunity, as they never had before, to indulge their tastes for novelties. They no longer address each other in the ordinary – *Comment cela va-t-il?* How goes it? But *Qu'as-tu mangé ce matin?* What had you to eat this morning?' [31].

Only by the beginning of January did some people finally realize that Paris was on the verge of starvation. Rationing of bread – the most basic staple of all – was instituted by the middle of the month, and the number on the relief rolls for food distribution quadrupled to 471,000 people [32].

At the beginning of the siege in September, Paris hospitals were reporting a total of some nine hundred deaths per week; by late-November, the number had risen to four thousand, and by January, to nearly forty-five hundred [33]. Smallpox, diarrhea, dysentery, pneumonia, and typhoid fever accounted for most deaths. Funeral processions with tiny coffins for infants and children became a common sight [34].

The siege had its greatest impact on mothers and infants, especially mothers who were breastfeeding. Émile Decaisne, a faculty member of the University of Paris, recorded his clinical observations of forty-three mothers and their infants. He conducted a careful chemical analysis of the mothers' breast milk and found its quality low. Most of the women had had little to eat in the weeks before he saw them. Their milk was relatively poor in fat and the protein, casein. Their infants, meanwhile, suffered from severe diarrhea and were, as Decaisne put it, 'dying from hunger'. After Decaisne's initial examinations and findings, when the women were given adequate amounts of food such as meat, bread, vegetables, butter, red wine, and coffee, the fat and casein in their breast milk increased substantially [35].

Among young children who were not breastfed, the lack of cow's milk during the siege added significantly to infant mortality [36]. Before the siege, Paris's normal daily milk consumption was eight hundred thousand liters, most of which was brought to the city from the provinces. Once the siege was under way, the city's daily milk supply dropped to a mere eighty thousand liters. Butter and eggs, too, became scarce, as these products could no longer reach Paris from the countryside. And because of their inflated costs, milk, butter, and eggs were impossibly beyond the reach of the poor.

Scientists therefore recommended that artificial milk be produced to feed babies and, presumably, reduce the death toll. The artificial milk concocted consisted of a watery emulsion of albumin, sugar, and fat, which some scientists thought were the basic elements of milk. To make matters worse, during the siege, the water supply of Paris had been severely compromised. The aqueducts – which brought potable water to the city – had been cut off, forcing the city to use mostly the Seine for drinking water. Much of the water that was distributed was unfiltered [37]. As infants continued to die, the famous French chemist, Jean-Baptiste Dumas, pointed out the insidious relationship among profit-seeking producers and sellers of 'the real equivalent of milk' and anxious mothers and vulnerable babies:

[Since no] conscientious chemist can assert that the analysis of milk has made known all the products necessary for life. . . we must renounce, for the present, the pretension to make milk. . . it is therefore always prudent to abstain from pronouncing upon the identity of these indefinite substances employed in the sustenance of life, in which the smallest and most insignificant traces of matter may prove to be not only efficacious, but even indispensible. . .. The siege of Paris will have proved that we. . . must still leave to nurses the mission of producing milk [38].

By the time the Siege of Paris ended on January 28, 1871, more than 50,000 civilians had died [39]. The following day, an armistice was signed at Versailles by Jules Favre, Vice-President of the Government for National Defense, and Bismarck.

The scarcity of vitamin A-rich foods undoubtedly contributed to the high mortality among infants and young children in Paris during the siege. This was hardly a unique episode – vitamin A deficiency occurred during the Siege of Malta (1798–1800) by the British during the Mediterranean campaign of 1798 [40], the Siege of Cádiz (1812–1814) by the French during the Peninsular War [41], and the Siege of Metz (1870) by the Prussians during the Franco-Prussian War [42].

Dumas hinted at the existence of vitamins, which were not known as the time, when he made the important declaration that there were yet unidentified 'substances employed in the sustenance of life in milk, and that these 'most insignificant traces of matter may prove to be not only efficacious, but even indispensible'.

References

1 Billard, C. (1828) Traité des maladies des enfans nouveax-nés et à la mamelle, fondé sur de nouvelles observations cliniques et d'anatomie pathologique, faites a l'Hôpital des Enfans-Trouvé de Paris, dans le service de M. Baron. Paris, J. B. Baillière.

2 Dupoux, A. (1958) Sur les pas de Monsieur Vincent. Trois cents ans d'histoire parisienne de l'enfance abandoné. Paris: Revue de l'Assistance Publique à Paris.

3 Another cause of ulceration of the eyes in infants was neonatal ophthalmia, a gonorrheal infection of the eyes passed on from their mothers, but, in contrast, this condition was characterized by a discharge of pus from the eyes.

4 Equivalent to a medical resident.

5 Semba, R. D., Victora, C. G. (2008) Low birth weight and neonatal mortality. In Semba, R. D., Bloem, M. W. (eds) Nutrition and health in developing countries. Second edition. Totowa, N. J., Humana Press, pp. 63–86.

6 Fuchs (1984), p. 143.

7 Billard (1828). Other descriptions of corneal ulceration in poorly fed infants were made around the same time; see Semba, R. D. (2007) Handbook of nutrition and ophthalmology. Totowa, N. J., Humana Press, p. 2.

8 Ollivier, [C. P.] (1832) Notice historique sur la vie et les travaux de C. M. Billard, docteur en médecine de la Faculté de Paris. Paris, J. B. Ballière.

9 Billard (1828), p. viii.

10 Billard (1828), p. viij.

11 Wet nurses employed in foundling homes in Paris produced about three pints (1.4 liters) of breast milk per day (see Budin, P. (1900) Le nourrisson. Alimentation et hygiène: enfants débiles, enfants nés à terme; leçons cliniques. Paris, Octava Doin). A resident wet nurse caring for 4–6 infants, with feedings divided evenly between the infants, would provide 0.23–0.35 liters of breast milk per infant, which is 30–45% of the usual milk intake. The average milk intake per day of healthy infants 0–6 months is 0.78 liters/day [Food and Nutrition Board (2001), p. 38]. According to the Institute of Medicine, Food and Nutrition Board, the Adequate Intake (AI) of vitamin A for infants 0–6 months should be 400 µg RAE/day [Food and Nutrition Board (2001), p. 111]. The vitamin A content of mature breast milk in wet nurses was probably in the range of 100–400 µg/l, which would be consistent with the quality of diet of the resident wet nurses. The resident wet nurses provided an estimated 23–140 µg RAE of vitamin A per day for infants in the Hospice, or about 6–35% of the AI for infants (0.23 l × 100 µg/l = 23 µg RAE/day low estimate; 0.35 l × 400 µg/l = 140 µg RAE/day high estimate).

12 Dupoux (1958), p. 305.

13 Pasteurization of milk was an innovation introduced in the 1880s: von Soxhlet, F. (1886) Über Kindermilch und Säuglings-Ernährung. Münchener medizinische Wochenschrift 33, 253, 276.

14 Frechier, D. M. (1840) Un mot sur une héméralopie qui règne épidémiquement dans le département des Bouches-du-Rhône. Bulletin général de thérapeutique médicale et chirurgicale 20, 248–249; Anon. (1855) Hemeralopie und Nyctalopie. Medicinische Zeitung Russlands 12, 361–363, 369–371; von Hecker, C. (1864) Klinik der Geburtskunde. Beobachtungen und Untersuchungen aus der Gebäranstalt zu München. Leipzig, Wilhelm Engelmann; Spengler (1865) Hemeralopie bei einer Schwangeren. Monatsschrift für Geburtskunde und Frauenkrankheiten 25, 61–63; Wachs (1867) Hemeralopie; dritte Niederkunft; Nachtblindheit zu Ende der Schwangerschaft bis in die ersten Tage des Wochenbettes. Monatsschrift für Geburtskunde und Frauenkrankheiten 30, 24–32; Küstner, O. (1875) Ein Fall von Hemeralopie in den letzten Wochen der Gravidität und den ersten Tagen des Wochenbetts. Berliner klinische Wochenschrift 12, 583–584; Kubli, T. (1887) Zur Lehre von der epidemischen Hemeralopie. Archiv für Augenheilkunde 17, 409–411; Walter, O. (1893) Ein Beitrag zur Lehre von epidemischen Nachtblindheit. Archiv für Augenheilkunde 28, 71–99; Cohn, S. (1890) Uterus und Auge. Eine Darstellung der Funktionen und Krankheiten des weiblichen Geschlechtsapparates in ihrem pathogenen Einfluss auf das Sehorgan. Wiesbaden, J. F. Bergmann.

15 Husson (1862), p. 311.

16 In addition to vitamin A deficiency, deficiencies in other micronutrients such as zinc and iron were likely present; vitamin A deficiency has the closest relationship among the micronutrients to higher infant mortality.

17 Wasting.

18 Dupoux (1958), p. 305.

19 Anon (1890) Pierre Bitot. Nécrologie. Mémoires et bulletins de la Société de Médecine et de Chirurgie de Bordeaux 1889, pp. 10–12.

20 Bitot, [P.] (1863) Mémoire sur une lésion conjonctivale non encore décrite, coïncidant avec l'héméralopie. Gazette hebdomadaire de médecine et de chirurgie 10, 284–288.

21 Mendes, J. C. (1862) Estudo sobre a hemeralopia á propósito dos casos observados na guarnição de Lisboa. Escholiaste medico 13, 22–24, 39–42, 55–58, 70–72, 85–88, 106–110.

22 Mecklenburg (1855) Hemeralopia epidemica im Gefängniss des Deutsch-Croner Kreisgerichts. Allgemeine medicinische Central-Zeitung 24, 73–75. For more details of these earlier descriptions, see Semba, R. D. (2007), pp. 3–4; de Hubbenet [von Huebbenet, A. C. A.]. (1860) Procès-Verbal de la Séance du 26 Septembre 1860 de la Société Médicale des Hôpitaux de Paris. Observations sur l'héméralopie. Bulletins et mémoires de la Societé Médicale des Hôpitaux de Paris 4, 559–561.

23 Postel-Vinay, G., Sahn, D. (2008) Explaining stunting in nineteenth century France. Cornell Food and Nutrition Policy Program Working Paper No. 199.

24 Horne, A. (1965) The fall of Paris: the siege and the Commune, 1870–71. New York, St. Martin's Press, p. 177.

25 Kranzberg, M. (1950) The siege of Paris, 1870–1871: a political and social history. Ithaca, Cornell University Press, p. 44.

26 Kranzberg (1950), p. 43.

27 Horne (1965), p. 182.

28 Horne (1965), p. 176.

29 Sibbet, R. L. (1892) The siege of Paris by an American eye-witness. Harrisburg, Meyers, pp. 328–329.

30 Horne (1965), p. 181.

31 Sibbet (1892), p. 33.

32 Kranzberg (1950), p. 137.

33 Sibbet (1892), p. 409.

34 Markheim, H. W. G. (1871) Inside Paris during the siege. London, Macmillan, p. 311.

35 Decaisne, E. (1871) Des modifications que subit le lait de femme par suite d'une alimentation insuffisante. Observations recueillies pendant le siége de Paris. Comptes rendus hebdomadaires des séances de l'Académie des Sciences 73, 128–131.

36 Horne (1965), p. 185; George Newman, an English public health specialist, claimed that during the Siege of Paris, 'while the general mortality was doubled, the infant mortality fell 40 per cent' owing to women's staying at home to breastfeed. Newman, G. (1906) Infant mortality: a social problem. London: Methuen and Co., p. 227. This claim is unsubstantiated and contrasts with the high number of infant deaths reported by the Paris hospitals.

37 Kranzberg (1950), p. 125.

38 Dumas, [J. B. A.] (1871) Note sur la constitution du lait et du sang. Le moniteur scientifique 3 ser., 1, 778–783.

39 Anon (1872) The mortality of the Siege of Paris. British Medical Journal ii, 131.

40 Robert (1803) Mémoire sur la topographie physique et médicale de Malte etc. In Mémoires sur l'Egypte, publiées dans les années VII, VIII, et IX. Tome quatrième, p. 30–111.

41 Deconihout (1834). Observations sur l'héméralopie, recueillies dans le mois d'avril 1833, à Mont-Dauphin, département des Hautes-Alpes. Recueil de memoires de médecine, de chirurgie, et de Pharmacie Militaires 36, 76–90.

42 Grellois, E. (1872) Histoire médicale du blocus de Metz. Paris, Ballière.

Chapter 4

Free but Not Equal

All persons held as slaves within any state or designated part of a state, the people whereof shall then be in rebellion against the United States, shall be then, thenceforward, and forever free. . ..
Such persons of suitable condition will be received into the armed service of the United States. . ..
President Abraham Lincoln in *The Emancipation Proclamation*, January 1, 1863 [1]

On May 31, 1897, Civil War veterans from the 54th Massachusetts Infantry marched in the rain across Boston Common past the newly unveiled memorial to Colonel Robert Gould Shaw and his regiment (fig. 4.1). The monument, a bas-relief by sculptor Augustus Saint-Gaudens erected more than three decades after the surrender of the rebel Confederacy, paid homage to the heroism and sacrifices of the Union Army's first black regiment. Eulogies by Mayor Josiah Quincy of Boston and Governor Roger Wolcott of Massachusetts put into words the gratitude of the United States for the deeds of its first black soldiers [2].

In fact, the fine words that preceded and followed the creation of the 54th Massachusetts Regiment revealed only part of the wartime circumstances of 1863. The establishment for the Union army of a 'colored regiment', as Governor Andrew of Massachusetts defined it, was as much a pragmatic measure as an idealistic gesture. Then in its third year, the Civil War was continuing to scorch the American landscape and torch its towns and cities, soaking the land with blood and strewing it with bodies clad in both blue and grey. The Confederacy was nowhere near defeat or surrender in 1863. In fact, it was about to inaugurate Jefferson Davis as its first president.

The military forces of both sides were being depleted, however, and recruitment was flagging. Both armies urgently needed new sources of manpower. While the South had enacted conscription in the Civil War's second year, the North had resisted this drastic, vociferously opposed measure. The one untapped source of volunteers for the North was able-bodied "men of color." In Massachusetts, the governor formed the 54th Regiment, and a vigorous multistate recruitment effort to staff it met with quick success. Black leaders addressed meetings to encourage volunteers. In Maryland, Frederick Douglass urged young men to sign up, and two of his sons did so. Black men from states as far afield as Illinois joined the volunteers from New England. As an inducement, the army offered each enlistee USD 13 a month pay, a

Fig. 4.1. Veterans of the 54th Massachusetts Infantry passing by the memorial to Colonel Robert G. Shaw and the members of the 54th Massachusetts Infantry, Boston Common, May 31, 1897. Courtesy of the Massachusetts Historical Society.

USD 100 bounty, plus 'Good Food and Clothing!' (fig. 4.2, 4.3). Far more men offered their services than were needed.

When initial recruitment was complete, the 1,007 black soldiers of the Massachusetts 54th would go to war under the supervision of forty white officers. Fully trained and armed with Enfield rifles, the unit was ready to mobilize by late May 1863. Cheers met the men as they marched through Boston past bunting-draped reviewing stands on the way to Battery Wharf to board the steamer, *De Molay*. Their ship was bound for the sea islands off South Carolina – that is, the entryway to the slaughterhouse that was the Confederate stronghold of Charleston.

In battle after battle, the soldiers of the 54th Massachusetts charged unflinchingly in the face of torrents of musket and cannon fire. When the fighting was reduced to hand to hand combat, they withstood the slashing of bayonets, swords, and handspikes. Their numbers diminished, as did those of the Rebel forces, but their skill, courage, and determination did not.

Instead of gratitude and reward, however, they garnered insults. Fear of black men bearing firearms engendered talk of disarming the soldiers of the Massachusetts 54th and equipping them just with pikes. And their pay, promised to equal to that of their white comrades – USD 13 a month plus a USD 3.50 clothing allowance – was deemed unwarranted. On July 2, 1863, Colonel Shaw wrote to Governor Andrew, 'You have probably seen the order from Washington which cuts down the pay of colored troops from USD 13 to USD 10. Of course if this affects Massachusetts

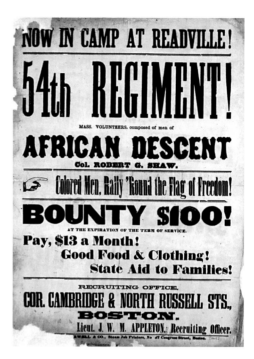

Fig. 4.2. Recruitment poster for the 54th Massaschusetts Infantry, 1863. Courtesy of the Massachusetts Historical Society.

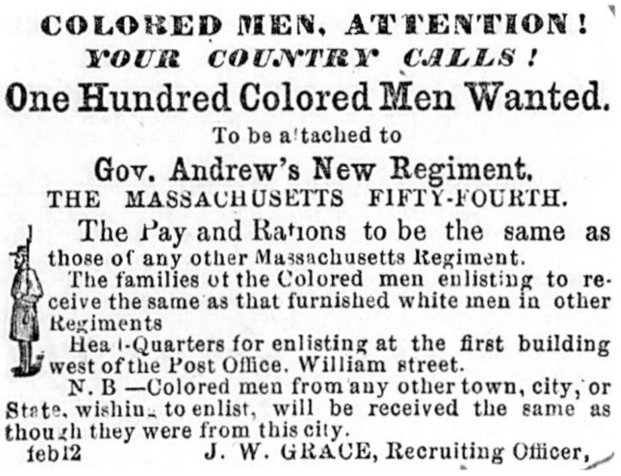
COLORED MEN, ATTENTION!

YOUR COUNTRY CALLS!

One Hundred Colored Men Wanted.

To be attached to

Gov. Andrew's New Regiment.

THE MASSACHUSETTS FIFTY-FOURTH.

The Pay and Rations to be the same as those of any other Massachusetts Regiment.

The families of the Colored men enlisting to receive the same as that furnished white men in other Regiments

Head-Quarters for enlisting at the first building west of the Post Office. William street.

N. B.—Colored men from any other town, city, or State, wishing to enlist, will be received the same as though they were from this city.

feb12 J. W. GRACE, Recruiting Officer,

Fig. 4.3. Advertisement from the New Bedford Mercury, February 1863. From the Norwood P. Hallowell papers, courtesy of the Massachusetts Historical Society.

regiments, it will be a great piece of injustice to them, as they were enlisted on the express understanding that they were to be on precisely the same footing as all other Massachusetts troops. . .' [3]. From the reduced USD 10, another USD 3 would be subtracted for clothing.

Rather than accept the arbitrary pay cut of almost 60%, which was exacerbated by reduced rations, the 54th Massachusetts regiment continued to fight without pay. It did so despite the crowning insult of a legal challenge to the men's legitimacy as soldiers. The federal government enacted a law in 1864 that limited military service for 'coloreds' to only those men who had been free on April 19, 1861, and administered an oath: 'You do solemnly swear that you owed no man unrequited labor on or before the 19th day of April, 1861. So help you God' [4]. The purpose, when a lump sum of back pay was offered to black soldiers, was to minimize the number of men who qualified – a transparent ploy at best. So all black men serving in the US forces swore in the affirmative and pressed on with their duties.

Finally, in September 1865, some twenty-eight months after its first mobilization, a much-reduced 54th Massachusetts was mustered out of service. Exactly how many men were present to be discharged is not known. Of the original one thousand seven, one hundred forty-one had died in action or from combat injuries, but a great many more had succumbed to illness. Of the regiment's known dead, infectious diseases, accidents, and unknown causes accounted for nearly two-thirds of its wartime deaths. Other black regiments experienced even higher mortality, especially from causes other than combat. The US 5th Colored Heavy Artillery regiment lost eight hundred twenty-one soldiers, of whom six hundred ninety-eight died from noncombat-related causes. Overall, 249,458 – nearly a 70% – of all 359,528 Civil War deaths of US troops were attributable to causes other than combat [5].

Race and Rank: Differences in Diet and Susceptibility

As the Civil War progressed, the number of soldiers who died with specific diseases – that is, the case-fatality rate – increased. For example, among soldiers who contracted typhoid fever between 1860 and 1864, the percentage of cases that proved fatal rose from about 18% to nearly 60% [6]. A pronounced imbalance between black soldiers and white emerges in the wartime disease-related fatality statistics. According to reports from the US Surgeon General's Office, white soldiers' noncombat-related deaths were markedly fewer than among their black counterparts (fig. 4.4) [7]. Black troops experienced higher incidence rates of smallpox, tuberculosis, mumps, diarrhea, measles, scurvy, anemia, and other diseases [8]. Steadily worsening malnutrition over the course of the Civil War has been implicated as the main factor underlying the increase in disease-related fatalities [9].

Night blindness was common in many regiments, but it occurred at a rate that was more than twice as high among black than among white soldiers (textbox 4–1). The number of soldiers affected with night blindness showed a clear peak in late spring and summer; in those seasons, foods containing vitamin A are relatively scarce. After the harvest season, vitamin A-rich fruits and vegetables are more readily available. Civil War troops' night blindness, therefore, gradually decreased from October through January. Moreover, the seasonal pattern of night blindness correlated closely with a prevalence of diarrhea – a related symptom of vitamin A deficiency (textbox 4–1; appendix).

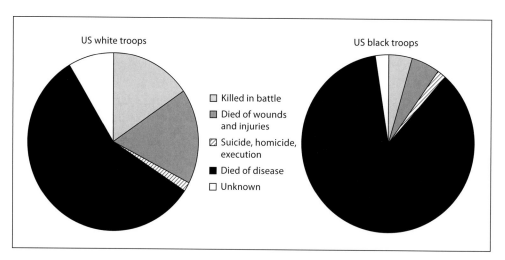

Fig. 4.4. Causes of death among Black and White US Troops, Civil War [7].

Textbox 4–1. Health disparities in the night blindness-diarrhea cycle

Diarrhea can worsen the effects of vitamin A in two ways. First, diarrhea impairs the intestinal absorption of vitamin A-rich foods. Second, a person suffering from diarrhea has an abnormal loss of vitamin A in his or her urine [10]. In turn, vitamin A deficiency can make the episodes of diarrhea more severe, since lack of vitamin A suppresses the immune system and the body's ability to fight the infection [11]. There was a seasonal pattern of acute and chronic diarrhea during the Civil War, with a peak of cases occurring in the summer.

The peak in cases of acute diarrhea and chronic diarrhea in the summer of 1864 coincided with a peak in cases of night blindness, and black soldiers were more affected than white soldiers by night blindness. This resulted in a heavy price for black soldiers – the underlying vitamin A deficiency undoubtedly contributed to a higher rate of deaths from acute diarrhea and chronic diarrhea in black compared with white soldiers (appendix).

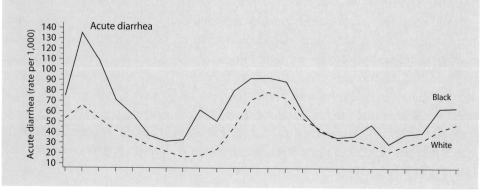

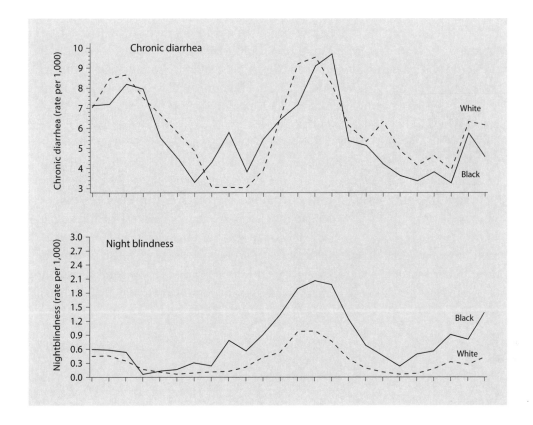

Given that vitamin A deficiency suppresses immunity, a higher prevalence of night blindness among black compared to white soldiers would be expected to be accompanied by high case fatality rates from the infectious diseases that can accompany night blindness. And indeed, the infectious disease case fatality rates in Union soldiers were much higher among black than among white soldiers (see fig. 4.5) [12]. The cause underlying the disparities is not immediately obvious. The quality of medical care delivered to black and white soldiers was considered roughly uniform, and black and white soldiers were hospitalized in the same facilities. The treatments for black and white patients suffering with infections were generally the same – i.e. ineffective (textbox 4–2) [13]. Underlying the difference in mortality was a fundamental difference in nutritional status.

Textbox 4–2. The liver conundrum

Night blindness was usually treated during the Civil War much as it was treated in nineteenth-century navies, namely with vesicatories and strychnine, and by enclosing the patients for a period of days or longer in complete darkness (chapter 1). Another course of treatment could have been more effective, but wartime circumstances made it impossible.

A different, empirical treatment handed down in medical writings from antiquity proved effective in curing night blindness and also concealed clues to prevention. It relied on animal liver. The organ would be roasted and fed to the patient; sometimes, the patient's eyes would be fumigated with smoke and steam arising from the roasting liver before he consumed it [14]. A version of the same cure involved giving the patient cod liver oil [15]. One physician extracted ox liver with alcohol and used the same alcohol (which now contained the extracted vitamin A) as an oral remedy to cure night blindness [16]. Night blindness could be cured within a few days of one of these liver regimens.

In hindsight, the reason for the effectiveness of the liver treatments is obvious. The liver is an animal's storage organ for vitamin A, thus, liver is the richest animal source there is of vitamin A. Some physicians for both the Union and Confederate armies were aware that ingestion of liver or cod liver oil was an effective cure for night blindness. But, ironically, they were prevented from following this course for the same reason that many soldiers came to them complaining of night blindness: scarcity.

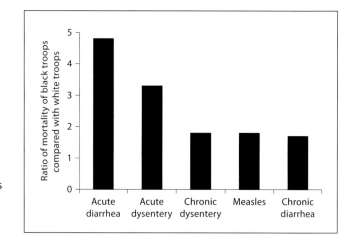

Fig. 4.5. Ratio of the case fatality rates of black compared to white troops. As an example below, black soldiers were nearly five times more likely to die from acute diarrhea than white soldiers [12].

The ideal of having a company cook and prepare decent meals for the Union troops broke down early in the war because, some people said, the men recruited to cook were 'uncouth and unqualified' [17]. As a result, the fighting men usually accepted whatever monotonous rations were available, adding whatever other food they could obtain, and prepared their meals in small messes of three or four soldiers. The only cooking equipment needed in an improvised mess was a frying pan and a pot to boil water for coffee. Cooking was a matter of frying a slab of bacon or salt pork, boiling coffee, and maybe taking some hardtack – a dense bread cracker roughly three inches square and a half-inch thick – and frying it up with the bacon or soaking it in the coffee. Hardtack was the Union soldiers' staple. Shipped in crates marked BC for Brigade Company, it was manufactured in Maryland and shipped to frontlines wherever they were. Soldiers wisecracked that the BC imprint meant the stuff had been made Before Christ. Nearly unbreakable while dry, hardtack had to be soaked before it could be bitten, and it often became ridden with weevils [18]. Many soldiers simply gave up on cooking and accustomed themselves to eating fat bacon or other meat, uncured and uncooked [19].

In a memoir, one Union soldier gave as complete as possible an inventory of army fare:

> I will now give a complete list of the rations. . . salt pork, fresh beef, salt beef, rarely ham or bacon, hard bread, soft bread, potatoes, an occasional onion, flour, beans, split peas, rice, dried apples, dried peaches, desiccated vegetables, coffee, tea, sugar, molasses, vinegar. . . pepper and salt. . .. [T]hese were not all served out at one time. There was but one kind of meat served at once. . . usually pork. When it was hard bread, it wasn't *soft* bread or flour, and when it was peas or beans it wasn't rice [20].

The marching ration for US troops was one pound of hard bread, three-fourths of a pound of salt pork or one and a quarter pound of fresh meat, sugar, coffee, and salt [21]. Rations rarely included dried fruits or vegetables. Anyway, the soldiers detested desiccated vegetables, calling them 'desecrated vegetables' or 'baled hay' [22]. When provided, the desiccated vegetables took the form of a bricklike cake of mixed vegetables that was so tough, to be edible, it required more than three hours' boiling in water – time that soldiers on the move rarely had [23].

Even not at its worst, the Civil War soldier's diet was almost devoid of vitamin A.

As was true of naval officers (chapter 1), commissioned Union officers were paid more and ate better than enlisted men. Officers' monthly pay scale was as follows (compared to white soldiers' monthly USD 13 and black soldiers' USD 10) [24]:

Colonel – USD 95
Lieutenant Colonel – USD 80
Major – USD 70
Captain – USD 60
First Lieutenant – USD 50
Second Lieutenant – USD 45

In addition, officers had – besides personal help – cash allowances with which they could acquire supplies from the brigade commissary. For infantry officers, the ration allowance and personal service scale was as follows [25]:

Colonel – six rations valued at USD 56 total and two servants
Lieutenant-Colonel – five rations valued at USD 45 and two servants
Major – four rations valued at USD 36 and two servants
Captains, First and Second Lieutenants – four rations valued at USD 36 and one servant

Fig. 4.6. Sutler's tent at Army of the Potomac Headquarters, Bealeton, Virginia, August 1863 (Photograph courtesy of the Library of Congress, American Memory Collection).

Civil War officers and enlisted men on both sides supplemented their rations with food from other sources. Sutlers, who were civilian tradesmen formally appointed to accompany the troops for the express purpose of selling food and other goods to soldiers, could augment army-issue rations and supplies (including weapons) – for a price. Their stocks, sold from tents pitched near the camps, included fruits, vegetables, butter, cheese, eggs, canned goods, and pastries. Because of the high prices they charged, much of what they sold was acquired for officers, who had the money (fig. 4.6). But the lower-paid soldiers eventually succumbed to the sutlers' offerings even if they had to buy on credit, which the sutlers willingly extended. On paydays, soldiers settled their debts to the sutlers by turning over up to four months' wages.

The cheaper way for Union troops to supplement army-issued rations was a practice the soldiers called 'foraging'. Defying orders, they stole what food they could find in houses, on farms, and in country stores; they also picked from fields and orchards. Even though the stolen property came from Confederate-owned sources, Union commanders did not authorize foraging, because, as one chronicler put it, 'prominent government and military officers thought that a display of force with consideration shown for the enemy's property would win the South back to her allegiance with the Union' [26]. As the war dragged on, however, Union officers tended to turn a blind eye to a soldier's appearing in camp with a pilfered chicken or turkey.

In the Confederate Army, which faced particular difficulties in supplying its troops, foraging became the de facto means of feeding soldiers [27]. But foraging eventually brought diminishing returns, as thousands of soldiers in both grey and blue traversed the same terrain. In some instances, food stores became depleted. In others, local owners learned either to hide their food supplies or feign destitution.

Of these two approaches to satisfying soldiers' craving for edible and sufficient food, patronizing sutlers offered the better nutrition. Besides being more appealing than hardtack and such, a sutler's wares were better in meeting the troops' need for supplementary foods containing vitamin A. But the expense of sutlers' offerings remained an obstacle especially for black soldiers. Black troops' substandard pay rendered them unable to acquire from sutlers such foods as butter and cheese, which, in turn, almost certainly contributed to vitamin A deficiency and to higher rates of night blindness and the accompanying conditions. The higher prevalence of vitamin A deficiency among black soldiers, with its effects in compromising the immune system, resulted in higher rates of sickness and death from infectious diseases.

Uneven Nutrition outside the Union Army

The medical care of Confederate troops was considerably less organized than that of the Union Army, and documentation of the health and nutritional status of Southern troops is comparably less abundant. Historians of medicine and re-enactors of the Civil War have amassed considerable information on injuries, treatments, and surgical operations, but retrospective scientific analysis is limited by a relative paucity of data. Conclusions about diseases and nutrition on the Confederate side are more a matter of extrapolation and speculation.

What cannot be contested is that, while Union soldiers ate poorly during the war, Confederate troops fared even worse. Confederate soldiers' official rations were similar to the rations for the US troops and included pork, bacon, or fresh meat, bread, flour, or corn meal, and dried peas or beans or rice, coffee, sugar, vinegar, and salt [28]. But what documentation there is of Confederate troops' nutrition reveals one pervasive and extreme feature: hunger. The food supply to the Confederate troops was erratic because of the poor infrastructure for bringing supplies to the front and the Union's blockading of southern ports. Hunger was a constant complaint among Confederate soldiers [29]. In one Confederate regiment, the soldiers were issued one-third rations, described as 'a little meete and corne meal not fit for your horses to eat'. One soldier wrote that he and his comrades were living on 'dry cornbread and a combination of spoilt bones and fat which the commissary calls bacon' [30].

In addition to an unreliable food supply and spoiled rations, the Confederate currency was unstable. Runaway inflation made the soldiers' pay nearly worthless. In addition, desertion – hence no pay whatever – became a pervasive problem.

The diet of families during the Civil War was not very different from that of soldiers. Faithful to the prevailing dietary dogma of the day that animal protein was the most beneficial of all food components (chapter 5), many households spent about one-third of their food budgets on beef or pork. The rest went for such staples as flour, sugar, coffee and tea, potatoes, butter, salt, pepper, vinegar, and dairy products – mainly milk [31]. This holds for civilians families that lived near military garrisons in which night blindness was epidemic: unlike the men in uniform, the families were unaffected by exceptional vision problems.

Given that soldiers and civilians had similar foods to eat, but that the only the rank and file – and never officers – suffered from widespread night blindness, the difference has to be ascribed to a better diet and the butter, milk, and cheese to which only the civilians and officers had access. In other words, civilians and officers and not soldiers had adequate food sources of vitamin A.

However paltry and poor the food to which Civil War soldiers had access, the dietary conditions of prisoners of war were far worse. By US regulation, Confederate prisoners were explicitly given less to eat than soldiers, and in some instances prisoners were simply starved to death. Point Lookout Prison Camp in Maryland, where Confederate soldiers and civilians were held captive, was notable as one of the worst prison camps in the North. To this day, outrage persists over the mistreatment of prisoners of war carried out at Point Lookout, where a monument to them still stands [32].

Inmates at Point Lookout were reported to have, among other health problems, the dilated pupils and corneal ulcers characteristic of night blindness, and some were completely blind [33]. Corneal ulceration and keratomalacia from severe vitamin A deficiency occur mostly in infants and young children (chapter 2, 3), but they are rare in adults. Their occurrence at Point Lookout certainly suggests conditions of extreme deprivation.

The ruination and bitterness left behind by the War Between the States is widely known. The total cost in lives continues to be debated; an approximate death toll is more than one million people, Union and Confederate soldiers and civilians combined. There is consensus that, of all military deaths, only about one-third was the result of combat, and that infectious diseases accounted for most of the rest. What may never be agreed on, much less proven, is how different the death toll might have been had the armies been provided with adequate nutrition. But for much of the war, the lowly Civil War infantryman's service was synonymous with hunger and, for many, susceptibility.

This was not lost on their commanding officers. Ulysses S. Grant wrote in his memoirs of his acceptance at Appomattox, Virginia, of the surrender of Robert E. Lee:

General Lee, after all was completed and before taking his leave, remarked that his army was in a very bad condition for want of food, and that they were without forage; that his men had been living for some days on parched corn exclusively, and that he would have to ask me for rations and forage. I told him 'certainly', and asked for how many men he wanted rations. His answer was 'about twenty-five thousand'; and I authorized him to send his own commissary and quartermaster to Appomattox Station, two or three miles away, where he could have, out of the trains we had stopped, all the provisions he wanted. As for forage, we had ourselves depended almost entirely upon the country for that [34].

References

1   Lincoln, A. (1896) Lincoln's inaugurals, the emancipation proclamation, etc. Boston, Directors of the Old South work. Old South Leaflets No. 11, pp. 13–15.

2   City of Boston (1897) Exercises at the dedication of the monument to Colonel Robert Gould Shaw and the fifty-fourth regiment of Massachusetts infantry, May 31, 1897. Boston, Municipal Printing Office, p. 61.

3   Emilio, L. F. (1894) A brave black regiment: history of the fifty-fourth regiment of Massachusetts volunteer infantry, 1863–1865. Second edition. Boston, Boston Book Company, p. 48.

4   Emilio (1894), pp. 220–221.

5   Fox, W. F. (1889) Regimental losses in the American Civil War (1861–1865). Albany, Albany Publishing Company, pp. 521–527. There were a total of about 560,000 deaths among U.S. and Confederate troops, of which about 30% were related to battle and 70% were due to malnutrition, infectious diseases, and other causes; see Bollett, A. J. (2002) Civil War medicine: challenges and triumphs. Tucson, Galen Press, p. 32. In contrast, during World War II, there were 300,560 US troop deaths, of which 97% were related to battle and 3% were due to other causes. A recent analysis by J. David Hacker suggests that the death toll may have been higher than Fox's estimate. See Hacker, J.D. (2011) A census-based count of the Civil War dead. Civil War History 57, 306–347.

6   Bollett (2002), p. 330. Typhoid fever is a life-threatening infection caused by the bacterium *Salmonella typhi* and spread by poor hygiene and sanitation.

7   US Surgeon General's Office (1870) The medical and surgical history of the war of the rebellion (1861–65). Prepared under the direction of the Surgeon General Joseph K. Barnes, United States Army. Vol. 1. Part 1. Washington, D.C., Government Printing Office, p. xxxvii.

8   Bollett (2002), p. 328.

9   Bollett (2002), p. 331.

10   Semba (2007), pp. 62–63.

11   Semba (2007), pp. 34–39.

12   US Surgeon General's Office (1870), pp. 637–711.

13   Bollett (2002), p. 326.

14   Robert, A (1802) Mémoire sur la topographie physique et médicale de Malte, suivi de l'histoire des maladies qui ont régné dans cette ville parmi les troupes françaises, sur la fin de l'an 6, et pendant les années 7 et 8. Paris, P. Didot l'Aîné; Quaglino, A. (1856) Studi pratici su l'emeralopia. Gazzetta medica italiana, Lombardia 4 ser., 1, 271–274, 280–284; Chiralt, V. (1858) Sobre el tratamiento de la hemeralopia. Siglo medico 5, 387; Fritsch-Lang (1873) Note sur l'héméralopie. Lyon médical 12, 565–571, citing Perrier.

15   Quaglino, A. (1867) Hemeralopia. Richmond Medical Journal 3, 44–50; Garau y Alemany, J. (1881) De la hemeralopía y su frecuencia en el soldado. Revista especial de oftalmologia, dermatologia, sifiliografia, y afecciones urinarias, año 4, 2:282–207, 339–353.

16   Torresini, M. (1858) Sopra i vapori di fegato di manzo nell'emeralopia. Gazzetta medica italiana Lombardia 4 series, 3, 45–46.

17   Robertson, J. I. (1988) Soldiers blue and gray. Columbia, University of South Carolina Press, pp. 65–66.

18   Billings, J. D. (1887) Hardtack and coffee, or the unwritten story of army life. Boston, George M. Smith & Co, pp. 115–116.

19   Robertson (1988), p. 65.

20   Billings (1887), pp. 110–111.

21   Billings (1887), p. 112. The marching ration was completely lacking in vitamin A.

22   Robertson (1988), p. 70.

23   US War Department (1880–1901) The war of the rebellion: a compilation of the officials records of the Union and Confederate armies. Series I, vol. V. Washington, D.C., Government Printing Office, p. 663–664, Letter from Charles S. Tripler to General R. B. Marcy, November 25, 1861.

24 US War Department, Pay Department, Exley, T. (1888) A compendium of the pay of the army from 1785 to 1888. Washington, US Government Printing Office, pp. 48–49.

25 Billings (1887), p. 113.

26 Billings (1887), p. 231.

27 Edwards, G. T. (2000) Rations C. S. A. In Heidler, D. S., Heidler, J. T. (eds) Encyclopedia of the American Civil War: a political, social, and military history, volume 4. New York: W. W. Norton & Co., pp. 1608–1610.

28 United States. Surgeon General's Office. (1888) The medical and surgical history of the war of the rebellion. Part III, Volume 1. Medical history. Washington, Government Printing Office, p. 67.

29 Robertson (1988), p. 72.

30 Mitchell, R. (1988) Civil War soldiers: their expectations and their experiences. New York, Viking, p. 166.

31 Cummings, R. O. (1941) The American and his food: a history of food habits in the United States. Chicago, University of Chicago Press, p. 77.

32 Page, D. (2000) Point Lookout Prison. In Heidler, D. S., Heidler, J. T. (eds) Encyclopedia of the American Civil War: a political, social, and military history. Volume 3. New York, W. W. Norton & Co., pp. 1534–1535.

33 Hays, W. (1866) Nyctalopia – night-blindness. Cincinnati Journal of Medicine 1, 315–316; Gillispie, J. M. (2008) Andersonvilles of the north: the myths and realities of northern treatment of Civil War confederate prisoners. Denton, Texas, University of North Texas Press.

34 Grant, U. S. (1886) Personal memoirs of U. S. Grant. New York, Charles L. Webster & Co, Vol. 2, pp. 494–495.

Chapter 5

The Long, Rocky Road to Understanding Vitamins

No animal can live upon a mixture of pure protein, fat, and carbohydrate, and even when the necessary inorganic material is carefully supplied the animal still cannot flourish. The animal body is adjusted to live either upon plant tissues or the tissues of other animals, and these contain countless substances other than the proteins, carbohydrates, and fats. . .. Scurvy and rickets are conditions so severe that they force themselves upon our attention; but many other nutritive errors affect the health of individuals to a degree most important to themselves, and some of them depend upon unsuspected dietetic factors. . .. I can do no more than hint at these matters, but I can assert that later developments of the science of dietetics will deal with factors highly complex and at present unknown.
From F. G. Hopkins, *The analyst and the medical man*, 1906.

When popular columnist Anna Steese Richardson published advice to expectant mothers in 1916, her counsel reflected the entrenched nutritional dogma of her day. 'Stoke the engine of your body with the right sort of coal', she wrote, 'Keep it clear of cinders and clinkers, cleanse it with pure water, renew the worn parts with rest. . . What is the right kind of coal? Food-stuffs classified according to their chemical properties. . . water, mineral matter, proteins, carbohydrates, and fats' [1]. From the late-nineteenth century well into twentieth, proteins, carbohydrates, fats, and minerals were believed to be the food components essential to build and sustain human life. The limitations of that dogma were beginning to come to light, but scientific understanding of food was far from complete. Knowledge of what constitutes adequate nutrition was still only partial, in effect, a puzzle with pieces scattered around the globe and some not yet even suspected.

Certainly, the understanding of food and nutrition lagged far behind the other life sciences, and by no means had it taken the great leaps made in other areas of human health. Unlike the other sciences, comprehension of the vitamins and their role in maintaining health would not be arrived at by great breakthroughs, but, rather, through small, incremental steps, with some conjectures, mistaken conclusions, disappointments, and refutations along the way.

Ultimately, it was laboratory scientists experimenting with animals, not physicians attempting to heal human patients suffering the effects of vitamin deficiencies, who succeeded in identifying and fitting together the pieces of the nutrition puzzle. Recognition of the vitamins and how they function in sustaining life and health was to come mainly through laboratory investigation using animal subjects. With animals, experimenters could control and manipulate conditions, when necessary making their subjects sick in ways that could not be done with humans. Most nutritional scientists conducted experiments with mice or rats, which were cheaper to acquire and maintain than larger animals and faster to show results. The small rodents could be had in large numbers at modest cost, had less variation among individuals, and enabled investigators to duplicate experiments to test the validity of results [2]. But even the breeding of rats for use as experimentation, which was essential in the study of vitamins, did not begin until after the turn of the twentieth century.

Moving Beyond Old Assumptions and Around New Certainties

The health-related science conducted in the decades before and just after 1900 was in many ways exceptionally fruitful. Studies in microbiology and the development of the germ theory, the work of Louis Pasteur in Paris and Robert Koch in Berlin, enabled the identification of the bacteria that cause anthrax, puerperal fever, cholera, plague, tuberculosis, leprosy, and many other diseases (textbox 5–1). Infectious microbes and toxins became widely implicated in the causation of disease, and antiseptic techniques were being developed. But certain common diseases, of which night blindness was but one, seemed not to conform to the germ theory. Although undeniably a revolutionary advance in many aspects of health care, the germ theory also created something of an obstacle to understanding vitamins and the ills caused by their lack.

> **Textbox 5–1.** Right and wrong applications of the germ theory
>
> During the 1854 London cholera epidemic, the physician John Snow (1813–1858) linked the cholera deaths to the use of contaminated water from the Broad Street pump in Soho. Snow persuaded the authorities to remove the handle from the Broad Street pump and was therefore credited with having averted further deaths [3]. But Snow's understanding of the spread of cholera was ahead of its time, as what that of physician William Budd, who believed the disease was caused by a specific living organism found in the gastrointestinal tract and spread from person-to-person through contaminated drinking water [4]. Both physicians were defying the prevailing notion of the day. The spread of diseases such as plague, measles, cholera, and malaria was attributed to miasmas rising from decaying organic matter.
>
> The concept that diseases are caused by infectious organisms or toxins produced by these organisms – that is, germ theory – supplanted the attribution of diseases to bad air by the late nineteenth century and became the reigning principle in pathology and public health. Physicians soon demonstrated, too, that diseases were spread by contagion, either by direct person-to-person contact or by mutual

contact with an intervening object or body (i.e. a vector). The French chemist Louis Pasteur (1822–1895) and the German physician-bacteriologist Robert Koch (1843–1910) further elaborated the germ theory of disease.

Germ theory gained acceptance in the late-nineteenth century and enabled the identification of the organisms responsible for anthrax [5], malaria [6], tuberculosis [7], cholera [8], leprosy [9], and diphtheria [10]. Given the productivity in determining the etiology of such diseases, the same line of inquiry into infectious diseases was often but mistakenly applied to the baffling diseases that in the early twentieth century were finally discovered to be caused by vitamin deficiencies. Such dead ends included the ideas that an intestinal infection caused scurvy, an unknown contagious disease caused rickets, and poor hygiene and sanitation facilitated an infectious disease that resulted in pellagra.

The growth of nutritional science in particular was stunted by accepted but insufficient theories. Many scientists in nutrition carried out their research within the given framework of food's having four basic components, and focused in particular on proteins and calories. Leaders in this work included chemist Justus von Liebig at the University of Giessen, physiologist Carl von Voit at the University of Munich, physiologist Max Rubner at the University of Berlin, and biochemist Russell Chittenden at Yale University.

As with assembling anything that is still fragmentary, some pieces of the nutrition puzzle appeared to fit but in fact did not. Liebig, for example, held that, besides red cells, the two principal components of the blood are the nitrogenous substances, albumin and fibrin [11]. When a person eats meat, Liebig contended, her or his digestive juices convert the nitrogen-bearing proteins into albumins, which, in turn, are absorbed as components of blood; the albumen and fibrin become muscle. (He acknowledged that plant foods also contain nitrogenous substances such as fibrin, albumin, and casein, and these too are absorbed and integrated into tissues much as are those in meat.) When muscles exert force, according to Liebig, muscle tissues are consumed and broken down into urea, carbonic acid, and water. Thus, muscles consume themselves during exercise, but the proteins absorbed from foods constantly replace them. During the day, according to this theory, muscle tissues break down; they are reconstituted at night during sleep. Liebig assigned to starch, sugar, and other non-nitrogenous food-borne compounds the role of supporting respiration. The only true nutrients, Liebig held – that is, substances that warrant designation as essential components of food – are the nitrogenous substances that can form or replace active tissue. Simply stated, Liebig's dogma identified proteins as the only source of muscular energy and physical activity as the only way to break down these substances. Liebig presented these ideas in 1842 in 'Animal Chemistry, or Organic Chemistry in Its Applications to Physiology and Pathology' [12].

Many scientists, but not all, accepted Liebig's clever scheme as a great step ahead in understanding nutrition. But one critic was quick to point out that the theory was inconsistent with real life, in which sedentary businessmen lived on meat-rich diets, while hardworking laborers carried out their physical toils while living on high-carbohydrate, protein-poor diets [13].

Looking back in the context of Liebig's theory at the night blindness suffered by sailors and soldiers (chapter 1, 4), the generally meat-rich military rations should have sufficed to render the men fit to carry out their duties, but clearly did not. The prevailing dietary dogma prevented physicians and scientific investigators from suspecting that night blindness was somehow related to a lack of something essential in servicemen's diets, and even though nutritional research was gaining prominence, a theory of vitamins that would account for the problematic missing elements was not even on the horizon. The notion that disease could result from dietary deficiencies also ran counter to both the prevailing ideas regarding the nature of food and the infection/toxin theory of disease. At the same time, however, clinical observations and scientific experimentation had begun to suggest that certain diseases did not originate with germs or toxins but with food.

The most widely known of these illnesses – though none was yet recognized as caused by a vitamin deficiency – were scurvy, beriberi, rickets, and pellagra. By the turn of the twentieth century, the clinical features of these diseases were well recognized, and to some degree, so were the circumstances under which they developed. The dietary changes that could cure or prevent scurvy and beriberi were increasingly adopted. The dietary deficiencies that caused them – vitamin C in the case of scurvy, thiamin in the case of beriberi, vitamin D in the case of rickets, and niacin in the case of pellagra – had yet to be determined.

The clinical picture of what resulted from vitamin A deficiency, in contrast, was more incomplete and fragmentary. It consisted mainly of descriptions of night blindness among sailors, soldiers, and children in orphanages, and, in some instances, awareness that diarrhea and dysentery could be associated with night blindness. Bitot's spots might be noted, also corneal ulceration, keratomalacia, and, ultimately, total blindness. And death rates among adults and children who presented these symptoms were extremely high. The only cures known, however, seemed to be ingestion of liver or cod liver oil, which was an empirical remedy familiar since antiquity – that is, it was known to work, but not how. The *cabinet tenebreux* also seemed to work in treating night blindness, but it was difficult to determine if a patient had truly been cured or was just tired of being shut up in dark closet for days on end. Any other remedy for night blindness, much less its cause, remained persistently elusive.

Characterizing vitamin A, from the initial glimmerings of its existence to its eventual isolation, purification and synthesis, was a long, much-interrupted process that, overall, took more than one hundred and thirty years. The 1816 dietary experiments of François Magendie in Paris, in which the investigator purposefully produced corneal ulceration in dogs by nearly starving them, yielded an early hint. In the next generation, Charles Billard, attending severely undernourished infant orphans with corneal ulcers, recognized the relevance of Magendie's observations and suspected a link between inadequate diet and corneal ulcers. But more than a half-century passed after Magendie and Billard's time before significant progress was made toward a theory of vitamins – specifically, important steps made in a laboratory in northern Europe's Baltic region.

Finally, studies were begun that forecast the intense, fast-paced period during which the vitamins finally came to be identified, characterized, and synthesized. Nicolai Ivanovich Lunin, a doctoral candidate studying chemistry in the laboratory of Gustav von Bunge (1844–1910) at the University of Dorpat in Estonia, showed in 1881 that adult mice could live in good health on milk. The mice in Lunin's experiments did not survive on a diet consisting of the milk components caseinogens (milk proteins), milk-fat, milk-sugar, and the ash of milk (i.e. proteins, fats, carbohydrates, salts, and water). In publishing his results, Lunin stated that, 'Mice can live quite well under these conditions when receiving suitable foods (e.g. milk), however, as the above experiments demonstrate that they are unable to live on proteins, fats, carbohydrates, salts, and water, it follows that *other substances indispensable for nutrition must be present in milk* [emphasis added] besides caseinogens, fat, lactose, and salts' [14]. Lunin is sometimes considered the first scientist to hypothesize that some uncharacterized substances essential for life were present in milk. French chemist Jean-Baptiste Dumas had made a somewhat similar observation during the Siege of Paris in 1871. It was von Bunge, however, and not Lunin (who had gone on to practice pediatrics), who pressed the matter.

Von Bunge reiterated the question in an influential textbook published in 1887: 'Does milk contain, in addition to (protein), fat, and carbohydrates, other organic substances, which are also indispensable to the maintenance of life?' [15]. A study by another von Bunge student, Carl A. Socin, of the different forms of iron in the diet, showed that mice fed only egg yolk (a rich source of vitamin A and iron), lived for nearly one hundred days, while mice fed an iron-poor diet with or without other forms of iron died within one month. Socin demonstrated that there was an unknown substance in egg yolk that was essential to life, and he raised the question of whether this substance was fat-like in nature [16]. Although von Bunge remarked, 'It will be useful to continue these investigations', neither he nor his students explored this promising new territory much further; the professor's main interest was the study of inorganic elements in nutrition.

At roughly the same time in England, Frederick Gowland Hopkins (1861–1947) played a key role in becoming one of the early scientists to experiment with feeding animals isolated food components. In time, Hopkins would attain the status of giant in the field of biochemistry and become a Nobel laureate. Originally trained as a chemical analyst, Hopkins began his professional life as a scientific toxicologist and built a reputation as an expert witness in poisoning homicides. He conducted chemical analyses for such celebrated cases as the trials of Adelaide Bartlett (husband died of chloroform poisoning), Florence Maybrick (husband died of arsenic poisoning), and Israel Lipski (neighbor died of nitric acid poisoning). Changing course quite early in his career, Hopkins went to the University of London and studied medicine at Guy's Hospital. At age thirty-seven, he was invited by the physiologist Michael Foster to teach and develop chemical physiology at Cambridge University.

Hopkins was interested in the chemistry of proteins, and at age forty in 1901 he isolated the amino acid, tryptophan (Glossary) [17]. The isolation of tryptophan was an extraordinary early achievement, made especially remarkable by the fact that, as Hopkins noted, he '. . .went to Cambridge without any training as a specialized biochemist' [18]. Unlike many colleagues of his generation, he had never spent time as a visitor in a German scientist's lab nor, for that matter, at the side of any master. Undeterred by his unconventional professional beginnings, Hopkins voiced objection to the idea, derived from classical cytology, that protoplasm was the living substance of cells. Many of Hopkins's colleagues believed that molecules of food and oxygen enter this mysterious complex, lose their identity there, and then emerge as recognizable substances such as urea and carbon dioxide. But taking a biochemist's view, Hopkins regarded the notion of protoplasm much as Pasteur and the other microbiologists regarded the obscure concept of spontaneous generation (i.e. the concept that life can arise from inanimate matter). Both protoplasm and spontaneous generation became increasingly improbable. Accordingly, Hopkins saw no reason why the chemical reactions inside a cell itself should be any different from chemical reactions observed in the laboratory [19]. He was also skeptical of the prevailing view that the food value of diets was based upon their energy and nitrogen values. 'Recent advances in physiology', he declared, 'seem to justify a fresh attack upon this subject (following) somewhat different lines' [19].

With colleague Edith Willcock, Hopkins fed carbohydrates, fats, and minerals to mice, thus demonstrating that tryptophan was an 'essential' amino acid. (The so-called 'essential' amino acids that are indispensable for life are those that the body cannot synthesize.) The only source of protein in the experimental diet was zein, the plant protein derived from maize that contains no tryptophan. Hopkins and Willcock's mice died unless tryptophan was added to the diet. Their study was the first to show that a specific amino acid was necessary, i.e. 'essential', in nutrition and led the way to identification of other essential amino acids [20].

A month before the 1906 publication of the Hopkins-Willcock amino acid findings, Hopkins, as Examiner in Pharmacology and Therapeutics at the Institute of Chemistry, spoke before the Society of Public Analysts at Burlington House in London. His lecture, titled 'The Analyst and the Medical Man', asserted that, 'the doctor and the analyst share between them. . . almost the entire burden of the maintenance of public health'. He then turned to the subject of dietetics, '. . .in which the medical man is the recognised authority, charged with instruction of the public, but for a scientific knowledge of which he depends largely on the chemical physiologist and the analyst'. He acknowledged that, as 'public analysts' – i.e. scientists – the members of the society played a role in protecting against the adulteration of foods and maintaining a safe food supply. He then presented new findings about the composition of food itself – facts that were '. . .less known and seemingly academic. I believe, however, that my theme, which is that the influence of minimal qualitative variations in dietaries, will one day become recognized as of great practical importance'.

Seeking an example, Hopkins reviewed the tryptophan study, in which he had found that 'a group in the protein molecule may serve some purpose in the body other than that of forming tissue or supplying energy'. Future analysts would be asked to do more than measure the total protein of a particular food; they would have to undertake the more difficult task of what Hopkins referred to as 'discriminative analysis'. His studies had led him to conclude that, '. . .no animal can live upon a mixture of pure protein, fat, and carbohydrate, and even when the necessary inorganic material is carefully supplied the animal still cannot flourish. The animal body is adjusted to live either upon plant tissues or the tissues of other animals, and these contain countless substances other than the proteins, carbohydrates, and fats'.

From there, he looked at the broad subject of diet and some of the widely known but not yet understood deficiencies:

> Physiological evolution, I believe, has made some of these well-nigh as essential as are the basal constituents of the diet. . . The field is almost unexplored; only is it certain that there are many minor factors in all diets of which the body takes account. . . In diseases such as rickets, and particularly in scurvy, we have had for long years knowledge of a dietetic factor; but though we know how to benefit these conditions empirically, the real errors in the diet are to this day quite obscure. They are, however, certainly of the kind which comprises these minimal qualitative factors that I am considering. . . Scurvy and rickets are conditions so severe that they force themselves upon our attention; but many other nutritive errors affect the health of individuals to a degree most important to themselves, and some of them depend upon *unsuspected dietetic factors*. . . I can do no more than hint at these matters, but I can assert that later developments of the science of dietetics will deal with factors highly complex and at present unknown [21].

The research underlying much of this address remained unpublished for a matter of years, during which illness and a severe injury caused a hiatus in Hopkins's career. Moreover, even as he recovered, the teaching responsibilities that came with his appointment as Science Tutor dominated his time. Shortly before he reached fifty, however, Hopkins was made a Fellow of Trinity College and Praelector in Biochemistry, a position once held by his mentor, Michael Foster. This post carried no teaching obligations, so he could finally work in this biochemistry laboratory without interruption.

In a critically important experiment carried out during this period, Hopkins observed that young rats grew poorly when fed a basal ration of protein, starch, cane sugar, lard, and minerals. With the addition of just a little milk to their diets, however, the animals grew normally. Hopkins published these remarkable findings in 1912 in the *Journal of Physiology* [22]. The unknown factors in milk that supported life were present in 'astonishingly small amounts'. A graph included in the paper shows the comparative growth patterns (fig. 5.1). After eighteen days of the experiment,

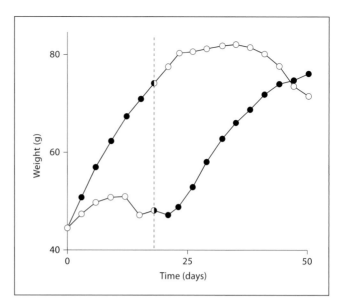

Fig. 5.1. Growth curves of mice in experiments of Frederick Gowland Hopkins (1912) [22]. Lower curve shows growth of 8 male rats with the basal diet, indicated by open circles (○). At day 18 (vertical broken line), the 8 rats were given 3 ml of milk per day, indicated by dark circles (●). Upper curve shows the growth of 8 similar rats receiving the basal diet plus milk. At day 18, these 8 rats were given the basal diet only with no additional milk.

Hopkins switched his subjects' diets so that the rats receiving the basal diet plus milk now received no supplementary milk; their growth subsequently stopped within about thirty days. Those receiving the basal diet alone began growing poorly within the first eighteen days, but after receiving milk each day starting on day eighteen, they grew rapidly. This study marks a critical early advance in the identification of a class of what Hopkins called food's 'accessory factors'.

Connecting the 'Accessory Factors' and the Vitamin Deficiency Diseases

Hopkins's work added significantly to the early evidence of the existence of vitamins and would eventually receive national and international recognition. But it was by no means specific or definitive. Nor did Hopkins establish whether there were many of these vital substances in food or just one. The biochemical nature of his 'accessory factors' and how they figure in human health was far from being determined. Efforts to understand a handful of familiar, debilitating diseases led investigators on paths toward the elusive substances in food that, when missing, could cause such trouble.

Beriberi
In the Netherlands, work parallel with Hopkins's was in progress in pursuit of a specific objective: how to prevent or cure beriberi. A result of thiamin deficiency, beriberi causes numbness or tingling in the extremities, difficulty walking sometimes to the point of leg paralysis, sometimes heart failure, and often death. In the late-nineteen century, beriberi was especially common in the Dutch East Indies, hence

a major concern for the Dutch government, which ruled a large portion of the area. The Dutch government formed a commission to study beriberi and appointed bacteriologist Cornelius Pekelharing (1848–1922) as its director. On the way to the East Indies, the commission stopped at Robert Koch's bacteriology laboratory in Berlin. The Dutch physician, Christiaan Eijkman (1858–1930), on sick leave after a severe episode of malaria, was augmenting his medical training in Koch's lab while recuperating. Pekelharing was impressed with Eijkman's talents and arranged for the younger physician to join the beriberi project.

The team's initial investigations – carried out in two rooms in a military hospital in Batavia (today's Jakarta) – involved autopsies of beriberi victims. The work included attempts to find suspected disease-causing bacteria in the cadavers' blood. Blood from beriberi victims was injected into various animals to determine whether the transferred blood would cause the experimental subjects to develop the disease. In the course of these studies, Eijkman observed that a polyneuritis, the equivalent of human beriberi, developed in chickens – specifically, in chickens fed grains of rice that had been polished of their bran sheathing ('silverskin') [23]. Eijkman concluded – incorrectly, but consistent with the dominant germ/toxin theories of disease the time – that the starch in polished rice carries a toxin, while the silverskin neutralizes the toxin or prevents its formation [24]. Only later did further findings from the Dutch East Indies prompt Eijkman to change his view on the causation of beriberi.

The beriberi problem, however, was not finally solved until a third, considerably younger, Dutch investigator, Gerrit Grijns (1865–1944), came to work in Java. Grijns continued the investigations begun by Pekelharing and Eijkman, but with an open mind. Grijns concluded – correctly – that beriberi was caused not by the *presence* of something pernicious but by the *absence* of something essential: the experimental chickens' diet lacked a vital component (textbox 5–2). Furthermore, Grijns observed – again, correctly – that the deficiency caused neurological damage. Publishing his findings in 1901, he wrote, 'There occur in various natural foods, substances, which cannot be absent without serious injury to the peripheral nervous system' [25]. Going further, Grijns observed that mung beans were potent in curing polyneuritis in chickens.

Textbox 5–2. Beriberi and 'water-soluble B'

As with night blindness and vitamin A, the correlation between beriberi and thiamin took many years to piece together. Nearly a half century passed between Eijkman's initial observations in Indonesia of a possible cause of beriberi and the identification and characterization of thiamin, a B complex vitamin – known during part of the interval by Elmer McCollum's term, 'water-soluble B.' In 1916, McCollum and a graduate student, Cornelia Kennedy (1881–1969), worked with pigeons to study polyneuritis (the avian form of beriberi); this disease was caused by lack of 'water-soluble B' [26].

In 1926, two Dutch chemists, Barend C. P. Jansen (1884–1962) and Willem F. Donath (1889–1957) succeeded in crystallizing 'water-soluble B' from rice bran-leftover from rice mills in Indonesia and identified it as thiamin [27]. The chemical structure of thiamin was later described and synthesized by Robert Williams (1886–1965) in 1936 at Columbia University and in the research laboratories of a New Jersey pharmaceutical company [28]. Fortification of flour with thiamin began in the 1940s in the US, and beriberi has become extremely rare since thiamin-fortified flour became widely available [29].

In 1902, another Dutch colleague, Dirk Johann Hulshoff-Pol, conducted a controlled trial with patients in a mental hospital in Java. With some three hundred patients housed in twelve pavilions, Hulshoff-Pol tested three different dietary interventions along with their regular polished rice diet. The patients were divided into four groups, which were subdivided for housing into three pavilions and maintained as follows:

The regular rice diet only (the control group),
The regular rice diet plus 300 grams per day of green vegetables,
The regular rice diet plus 150 grams per day of mung beans, and
The regular rice diet only, plus periodic disinfection

After nine months, none of the patients who had received mung beans developed beriberi, but 33% of the control group and 42% of the added-green-vegetables and disinfection groups did develop the disease. The addition of regular disinfectant fumigations was to test whether the disease was actually carried by cockroaches. The results showed that cockroaches were not to blame. The addition of mung beans to the diet effectively prevented beriberi [30].

Other dietary human trials were soon to follow in southeast Asia, including a notable experiment in 1909 conducted with two road crews in a remote rainforest on the Malay Peninsula [31]. Laborers eating white rice developed beriberi; those who ate brown rice did not. The relationship between new cases of beriberi and the type of rice consumed was consistent when the diets of two road crews were switched.

Rickets

Rickets, a common condition in children in Northern Europe and northern North America, was another disease that challenged the infection/toxin theory. A result of abnormally low mineralization and mechanical strength in developing bones, rickets is caused by lack of vitamin D; vitamin D is found in oily fish, in egg yolk, and, like vitamin A, in cod liver oil. Deformities such as bowlegs and knock-knees in the weight-bearing long bones are characteristic of rickets in children. Vitamin D is also generated in the skin through direct exposure to sunlight. Thus, both a deficient diet and lack of exposure to sunlight can contribute to rickets.

Observations made in a London zoo suggested that rickets might be caused by a dietary deficiency. In 1889, John Bland-Sutton, a surgeon at Middlesex Hospital and prosector at London's Regents Park Zoological Garden, reported on studies of rickets in monkeys, lions, and other animals [32]. The lion cubs had the classic bone deformities of rickets. Before an effective treatment was found, the misshapen cubs were removed from the exhibits and kept out of public view. Bland-Sutton surmised at first that the rickets was caused by a deficiency of fat in the diet. He recommended that, in addition to their usual diet of lean meat, the lion cubs be fed cod liver oil and crushed bones. The results were dramatic: the prescribed diet cured the rickets.

Bland-Sutton's findings were consistent with the empirical observations of the many pediatricians who for some time had been using cod liver oil to treat rickets in children [33]. In the early 1880s, Max Kassowitz, a pediatrician in Vienna, had introduced cod liver oil and phosphorus called 'Kassowitz formula' for the treatment of rickets in children [34].

Scurvy

Although long known to be preventable by including lemons and other specific foods in the diet, scurvy remained an enigma from a scientific standpoint – i.e. that it resulted from vitamin C deficiency – well into the twentieth century. A host of symptoms can be associated with scurvy, including softening of gums and loss of teeth, hemorrhages in the skin, weakness, and lower extremity edema. In 1907, two Norwegian scientists at the University of Cristiania (Oslo), bacteriologist Axel Holst and pediatrician Theodor Frölich, developed an animal model that enabled a breakthrough in the study of scurvy [35]. Holst had studied with Koch in Berlin and visited the Grijns laboratory in Batavia to gain insights from the beriberi studies in progress there. When he returned to Oslo, however, Holst was frustrated in his attempts to produce scurvy in chickens and pigeons, and he decided to work with Frölich using guinea pigs as experimental subjects.

The guinea pig was a fortuitous choice, since guinea pigs, like humans (and unlike dogs, rats, mice, and birds), require vitamin C as an essential nutrient. The use of the guinea pig was a departure from the usual practice in German and French laboratories of using dogs for feeding studies. However, the guinea pig was known as a children's pet in the early 1900s, and its space and food requirements were minimal compared with dogs' [36]. Holst and Frölich showed that guinea pigs fed grain, groats (the hulled grains of cereals), and bread developed scurvy, but those for which fresh cabbage or fresh potatoes were also provided did not. After reviewing the epidemiological and clinical data on scurvy in humans and their own experimental results, Holst and Frölich concluded: '. . .epidemiological facts speak in favor of the opinion that the described disease in guinea-pigs is identical with human scurvy'.

Night Blindness

The piecemeal clinical picture of night blindness caused by of vitamin A deficiency (see previous chapters) finally came together between 1896 and 1904, when Japanese physician Masamichi Mori described more than fifteen hundred children with *hikan*

– that is, xerophthalmia [37]. Mori had studied medicine at the Mie Prefectural Medical School and the Tokyo University and gone on to work in Germany and Switzerland before returning to Mie Prefecture to practice surgery. The children Mori described had night blindness, Bitot's spots, corneal ulceration, keratomalacia, and diarrhea. The death rate among them was high. Most were between ages one and four and one-half, and many came from poor families living in mountainous regions, where the diet completely lacked milk and fish. Once under medical care, the children were given cod liver oil daily, and this proved to be an effective treatment for both the eye lesions and diarrhea. Contrary to the view of many physicians, Mori concluded that the disease was not infectious but rather was caused by the lack of fat in the diet.

Pellagra

Pellagra, a disease characterized by skin lesions, diarrhea, wasting, neurologic and psychiatric disturbances, and high mortality, was known in parts of Europe. Pellagra is caused by a lack of niacin in the diet. Foods that are rich in niacin include red meat, liver, fish, poultry, legumes and cheese. The appearance of pellagra coincided with the cultivation of maize, and it became a common problem in southern and western France by the early nineteenth century. Persons afflicted with pellagra were mostly poor peasants who ate primarily corn, since they could not afford a more diverse diet. Physicians advanced several theories to explain the causation of pellagra, including the consumption of too much maize, an intoxication caused by eating moldy maize, and, following the doctrines of Liebig, that corn lacked sufficient protein [38]. Although the exact cause of pellagra was not identified at the time, the physician Théophile Roussel advocated changes such as increased animal husbandry and substitution of other cereal crops for corn that eventually led to the disappearance of pellagra from France [39]. By the time pellagra was largely forgotten in Europe, it made a new appearance among poor sharecroppers in the southern United States early in the twentieth century (detailed below).

Finding an Elusive Panacea in Milk

Polished rice and beriberi cured with whole grains in southeast Asia, rickety London lions and Japanese babies with dry, ulcerated eyes cured with cod liver oil, Norwegian guinea pigs cured of scurvy with cabbage – all these findings became familiar to European and American life scientists in the early 1900s (table 5.1). The dogma of only four essential food components was losing its grip on investigators interested in nutrition and health. The infection/toxin theory of disease, too, was beginning to seem inadequate to explain all illnesses. Nutrition scientists were warming up for the race to characterize the elusive food components that Hopkins grouped as 'accessory factors'. Wilhelm Stepp in Strasbourg, Thomas Osborne and Lafayette Mendel at Yale, and Elmer McCollum (ultimately) at Johns Hopkins University, Marguerite Davis, and Harry Steenbock at the University of Wisconsin – all of them went to work

Table 5.1. Observations of deficiency diseases by 1910

Disease	Characteristics	Clinical observations	Animal observations
Scurvy	bleeding gums hemorrhages in skin high mortality	cured or prevented with lemons, oranges, onions, green leafy vegetables	experimental scurvy produced in guinea pigs; disease prevented by adding cabbage to diet
Beriberi	numbness, tingling in extremities difficulty walking high mortality	cured or prevented with milk, brown (unpolished) rice, mung beans	experimental beriberi produced in chickens and pigeons (avian polyneuritis); disease prevented by adding brown rice to diet
Rickets	bone deformities	associated with bad hygienic conditions and lack of exercise	rickets in lion cubs in London zoo cured with cod liver oil
Xerophthalmia	Bitot's spots corneal ulceration blindness high mortality	cured or prevented with cod liver oil	conjunctivitis and corneal ulceration in rats fed a limited diet

feeding isolated food substances to assorted animals in hope of identifying what, for a while, came to be called 'vitamines' (textbox 5–3).

Textbox 5–3. A family name for the 'accessory factors'

With scientists in far-flung academic centers on both sides of the Atlantic closing in on F. G. Hopkins's 'accessory factors,' the still-elusive quarry needed a collective name. In 1912, Polish-born biochemist Casimir Funk proposed 'vitamine', from the Latin *vita* for life and 'amine', referring to the presumed chemical nature of the substances [40]. (Amines are a class of nitrogen-containing organic compounds derived from ammonia [NH_3]; the term provides the basis for the phrase 'amino acid'). McCollum had suggested a nomenclature of 'fat-soluble A' and 'water-soluble B'. In 1920, Jack Drummond, a biochemist and vitamin researcher at University College, London, proposed the term be shortened to vitamin. From the strict standpoint of chemistry, not all the 'vitamines' were chemical amines. Drummond also suggested dropping McCollum's designation of 'fat-soluble A' and 'water-soluble B'. Instead, he recommended, 'the substances should be spoken of as vitamin A, B, C, etc. This simplified scheme should be quite sufficient until such time as the factors as isolated, and their true nature identified' [41]. The new nomenclature, when Drummond proposed it and ever since, has been widely adopted in the scientific community and beyond.

A young medical graduate working in chemistry in Germany, Wilhelm Stepp (1882–1964), conducted studies that suggested the existence of fat-soluble substances essential for life. He mixed flour with milk and formed it into dough – he called it milk-bread – and fed it to experimental baby mice, which grew normally to adulthood. But when Stepp extracted any fats present in his milk-bread using alcohol and ether and fed the resulting paste to mice, the animals could not survive beyond three weeks. When the extracted substance was added back to the fat-free paste, the mice survived normally. Stepp's conclusion, published in 1909: there is some fat-soluble substance that is essential for survival [42]. In further studies published two years later, Stepp noted that 'certain lipid substances present in milk (that are) soluble in alcohol-ether, are indispensable for the survival of mice' [43]. After these valuable findings, Stepp turned his attention to the practice of medicine.

As a new generation of life scientists was reaching maturity, the center of biochemical research was undergoing a geographic shift, from Europe to the United States. Thus, one of the most influential early associations in biochemistry was housed not in France, Germany, or Great Britain but in the United States. Starting in 1909, Thomas Osborne (1859–1929) and Lafayette Mendel (1872–1935) – one from a socially prominent New England family, the other born to German Jewish immigrants in New York State – had a long, highly productive scientific collaboration in New Haven, Connecticut. Their work together yielded more than one hundred published papers.

Osborne had a long association with Yale, where he did his undergraduate and graduate studies and worked as an instructor. Osborne soon joined the staff of the Connecticut Agricultural Experiment Station, which was located about a mile north of the Yale campus. He initiated a project to investigate the proteins in plant seeds. His early work refuted a half-century-old thesis of Justus von Liebig, that four forms of plant protein – vegetable albumin, plant gelatin, legumin or casein, and plant fibrin – were identical to four animal proteins with similar names [44]. To analyze the proteins of different seeds, Osborne applied chemical techniques to determine the seeds' amino acid composition. His conclusion: the proteins of seeds are specific substances with distinguishing amino acids, and even among closely allied species, seed proteins have pronounced differences.

Osborne's next question: how influential are the different proteins in nutrition? Osborne would assess the relative nutritive properties of proteins in partnership with his friend, Lafayette Mendel. Mendel's lab was housed in the Sheffield-Town mansion, an Italianate structure with towers and cupolas, in which the old drawing room served as a lecture hall and the art gallery a large, somewhat makeshift laboratory. Here Mendel, a devoted and demanding teacher, supervised graduate students. They, in turn, kept their experimental animals in various rooms (fig. 5.2). In time, the laboratory sprawled into the mansion's tower and crannies in the basement [45].

Osborne and Mendel conducted feeding experiments with rats at the Connecticut Agricultural Experimental Station to determine the nutritive value of purified

Fig. 5.2. Thomas Osborne (standing behind skeleton) and Lafayette Mendel (standing, third from right, next to Osborne), with faculty and students, on the steps of the Sheffield-Town mansion, which served as the laboratory for the Department of Physiological Chemistry, in 1899. Photo courtesy of Yale University Archives and Manuscripts.

proteins. After some initial frustrations, including a fire that destroyed their lab in 1910, they developed for their rats a basal diet of isolated food substances consisting of casein, sugar, starch, lard, and minerals. With these elements combined, this diet served quite well to support life in rats, and it enabled the investigators to study and compare the effects of adding various seed proteins. Initial experiments revealed difficulty in maintaining adult rats for longer than about two to three months either on the basal diet or on diets to which seed proteins were added. Powder made from dried milk produced the distinctive results of maintaining the health of young rats, but when the milk powder was replaced with casein and mineral salts, the rats died [46]. The team then developed protein-free milk by removing the casein and other proteins, filtering the resulting solution, and drying it. This protein-free milk powder became an important component of their initial studies, which they summarized in the 1911 monograph, 'Feeding Experiments with Isolated Food-Substances' [2]. They continued their studies, acknowledging in 1912, '…with respect to the actual requirements of fat on the part of the healthy organisms there is at present almost no definite information available. Fats are, of course, commonly found present in greater or lesser abundance in every dietary; but to what extent they represent an indispensable need of the animal remains to be learned' [47].

This gap was soon to be filled not only by work in Osborne and Mendel's own laboratory but also from the lab of Elmer McCollum (1879–1967). A native of Kansas, where he did his undergraduate work, McCollum, too, pursued the PhD at Yale, where he studied organic chemistry. After completing his doctoral thesis in 1906 on the chemistry of the pyrimidines, McCollum paid a call on the man he described as 'the outstanding teacher and investigator in physiological chemistry in America', Lafayette Mendel [48]. As no academic positions were open, Mendel invited McCollum to spend the next academic year as a post-doctoral student in his laboratory. McCollum made the most of that year, attending lectures in biochemistry and learning new analytic methods. At the end of McCollum's post-doc year, Mendel arranged for his protégé to work in the Department of Agricultural Chemistry at the University of Wisconsin. McCollum's new mentor would be biochemist Edwin B. Hart (1874–1953).

At the time, the generally accepted method for analyzing food, devised by German scientists, was based strictly on chemical analysis of the food. Hart and his colleagues were skeptical of this approach. Instead, they sought to evaluate the foods by actually assessing how the foods affected animal growth. Hart's main project consisted of testing on cows single-item versus multiple-item diets, specifically, either maize, wheat, or oats, or a mixture of the three. According the German system of food analysis, these four diets were chemically equivalent. McCollum's first assignment was to record the cows' food intake and then analyze their milk and urine. In his second year in Hart's 'single grain ration experiment' McCollum was joined in his duties by Harry Steenbock (1886–1967).

Like McCollum a farm boy, Harry Steenbock had grown up in Wisconsin, where he did his undergraduate work at the Wisconsin College of Agriculture in Madison. Before graduating in 1908, Steenbock was one of just two students to attend McCollum's first lectures at the University of Wisconsin on the biochemical aspects of nutrition. McCollum and Steenbock spent many hours preparing large quantities of different feeds for Hart's cows. The different diets produced dramatically different results, McCollum recalled decades later in his autobiography. 'The wheat-fed cows were small of girth and rough-coated. They were all blind, as shown by the lead color of the eyes and by their inability to find their way about. . . (T)he corn-fed cows were, by standards of animal husbandry, in excellent condition' [49].

During this time, McCollum followed developments in animal and plant chemistry in Europe by reading the 'Yearbook on the Progress of Animal Chemistry' *(Jahres-Bericht über die Fortschritte der Tier-Chemie)*. Thus, he knew of the experiments of Hopkins, Lunin, Socin, Stepp, and Pekelharing, which involved feeding different diets to mice or rats to determine what constituted the simplest diet. This gave McCollum the idea of experimenting with rats instead of cows [50]. With his own six dollars, he purchased a dozen albino rats and started a breeding colony in 1908. His colleagues at the agricultural experiment station in Wisconsin frowned on the notion of raising rats, which were universally considered pests. But McCollum persisted. McCollum's early attempts to feed rats isolated and purified food substances were, in his words, 'of a bungling nature', because he had difficulty ascertaining the purity of the isolated dietary components [51]. At first, he thought the rats failed to thrive because the rations were not palatable [52]. But the care and feeding of McCollum's experimental rats began to change with the arrival of Marguerite Davis. Born in Racine, Davis had gone to California for her undergraduate work, but after graduating, she returned to Wisconsin. In July 1909, Davis presented herself at McCollum's lab and requested to study biochemistry there. McCollum was relieved to turn over to Davis the management of the rat experiments, as he was mostly occupied with Prof. Hart's single-grain-ration cow studies.

Hart's project ended, inconclusively, in 1911 [53]. Why the corn diet was so superior to other diets for the cows' health and growth remained unexplained. But to continue further work with cows, given their breeding pattern and lifespan, would probably have taken another four years. While Hart's experiment was winding down, McCollum's other main associate, Steenbock, also ceased to figure importantly in McCollum's experimental work. Steenbock left Wisconsin temporarily to seek further training in New Haven and Berlin. Only later, in 1914, would Steenbock return to Madison to complete his doctoral studies.

Although McCollum continued with Davis to pursue his investigations of isolated food substances in rats, his only publications in 1911 and 1912 were related to nutritional studies conducted with Hart and other colleagues using cows, pigs, and hens. In July 1913, McCollum and Davis reported in the *Journal of Biological Chemistry* that different fats were not equal in value for rats' growth – a finding that ran counter to the prevailing nutritional dogma that diverse fats had similar nutritional values. On a diet of casein, lard, lactose, starch, and salt, young rats grew normally so long as an ether extract from butter or egg yolk was added. (The ether extract was prepared by mixing ether with butter or egg yolk to dissolve out the fat-soluble substances.) The rats died, however, if a similar ether extract from lard or olive oil were added. The investigators concluded that, 'Our observation that ether extracts from certain sources improve the condition of animals on such rations, strongly supports the belief that there are certain accessory articles in certain food-stuffs which are essential for normal growth for extended periods' [54]. The differences in rats fed butterfat and those not fed butterfat could be dramatic (fig. 5.3). Many years hence, when looking back at this experiment, McCollum somewhat magnified the significance of its findings (textbox 5–4).

Textbox 5–4. Discovery and a Distorting Mirror Of Hindsight

In much later writings, Elmer McCollum stated that he had discovered vitamin A with his 1913 study [54]. He also maintained that, 'this observation was promptly verified by Osborne and Mendel' [55]. McCollum based this claim on his observation that the unidentified factor, which only later was identified as vitamin A, was fat-soluble. Thus, McCollum's responsibility for the identification of vitamin A was, to say the least, an exaggeration [56].

Socin (in 1891) and Stepp (in 1911) had long since suggested or demonstrated the fat-soluble nature of this unknown substance. The fat-soluble substance in butter and egg yolk in their experiments actually contained three vitamins: A, D, and E – three of the many 'accessory factors' that were far from identification in 1913. More incremental steps over many years had to be taken before these vitamins were separated, the chemical structure of vitamin A was described, and vitamin A was isolated, purified, crystallized, and synthesized. There is little justification to a claim that any single one of these many steps represented 'The Discovery of Vitamin A'.

The august Nobel committee in Stockholm exercised the caution of specificity when it used the word 'discovery' in awarding its 1929 Prize in Physiology or Medicine to Frederick Gowland Hopkins 'for his discovery of the growth-stimulating vitamins' and his Dutch colleague, Christiaan Eijkman, 'for his discovery of the antineuritic vitamin' [57]. Apparently more inclined than McCollum to give credit where it was due, Hopkins acknowledged that the 1881 study of Nicolai Ivanovich Lunin had implied the existence of the 'accessory factors'. He also paid respect to Cornelis Pekelharing, who in 1905 had conducted studies in Utrecht showing that milk contained unknown 'essential substances'. Pekelharing's experiments, though definitely relevant and earlier than Hopkins's, won little recognition outside the Netherlands in his own time because he published his work in Dutch, as did Gerrit Grijns [58].

Later, in 1937, when the Nobel Committee presented the Swiss chemist Paul Karrer with the Nobel Prize in Chemistry for describing the chemical structure of vitamin A, they further clarified the awarding of the 1929 Nobel Prize: 'The other half of the Nobel Prize in Medicine was awarded in the same year to Hopkins in recognition of his discovery of the vitamins of growth, that is, the substances necessary for the growth of the animal body – contained for instance in milk – and of which one of the most important has now been identified with vitamin A' [59]. This declaration left no doubt that the Nobel Committee had firmly closed the door to any other nominations for the 'discovery of vitamin A'.

McCollum's colleagues in New Haven, meanwhile, were coming to similar conclusions. In the August 1913 issue of the *Journal of Biological Chemistry*, Osborne and Mendel reported that rats fed a basal diet of isolated proteins, starch, lard, and milk with all its protein extracted grew normally for about sixty days, but then they declined and died. The addition of butter or replacement of lard with butter in the diet allowed normal growth in young rats. Osborne and Mendel concluded, 'In seeking for the "essential" accessory factor we have, therefore, been led first to supply the cream component, in the form of butter. . . (I)t would seem, therefore, as if a substance exerting a marked influence upon growth were present in butter. . .' [60].

Fig. 5.3. Two rats were identical at weaning time and were raised on the same basal diet, but the one on the left received sunflower seed oil (no vitamin A) and the one on the right received butterfat (rich in vitamin A). From McCollum, E. V. (1918) The Newer Knowledge of Nutrition. New York, MacMillan.

In the next year, Osborne and Mendel reported results from tracking down a clue that had surfaced several years before in Basel, Switzerland. In 1909, ophthalmologist Paul Knapp had reported that rats fed 'purified' diets of protein, carbohydrates, fats, and minerals not only grew poorly but also showed increased susceptibility to infection and eye problems. Specifically, Knapp's rats on the basal diet developed conjunctivitis and corneal ulceration, and before long, they died. When milk was provided, however, the subjects' eye disease and deaths were averted, leading Knapp to surmise that something in milk might be preventing eye lesions [61]. In 1914, Osborne and Mendel tested Knapp's proposition about milk. Specifically, they showed that rats on their basal diet developed inflamed eyes and diarrhea, but they recovered when cod liver oil or butterfat was added to their diet [62].

Meanwhile, Elmer McCollum too, arrived at similar conclusions. In 1917, he noted that, 'blindness results if the animals are permitted to go without this dietary essential or with an inadequate supply for a sufficient time' [63]. The elusive 'accessory factors' were sufficiently close to being identified that McCollum started identifying them with letters of the alphabet (textbox 5–3). In the following year, he wrote, 'Xerophthalmia and polyneuritis are abundantly demonstrated to have their origin in the lack of a sufficient amount of the fat-soluble A and water-soluble B, respectively, in the diet' [64].

The eye disease that civilian physicians saw in malnourished infants and children, that military doctors described in servicemen, and that scientists were inducing in experimental animal subjects proved to be the same disease. Its cause was lack of what McCollum termed 'fat-soluble A', which also became known as the 'anti-xerophthalmia vitamin'.

Obstructions, Chicanery, and Perseverance

With F. G. Hopkins' 'accessory factors' still unproven, diehard proponents of the four-basic-components food dogma stood their ground, pointing to possible flaws in the experimentation being done in the pursuit of accessory factors. A particularly acerbic voice was to be heard from Breslau, Germany. Franz Röhmann (1856–1919), a prominent investigator and staunch defender of Justus von Liebig, asserted, 'The assumption that some unknown substances are indispensable for growth is a convenient device for explaining experiments that result in failure – a device that becomes superfluous as soon as the experiment succeeds' [65]. Röhmann illustrated his case: he fed mice what he called a purified diet of proteins, carbohydrates, fats, and minerals, and not only did the mice grow to maturity, but also they produced healthy young [66]. In 1917, however, Osborne and Mendel discredited Röhmann's conclusion by noting that the commercial casein preparation Röhmann had used was impure and contaminated with other vitamins [67].

Four major problems did indeed muddy experimentation with animal subjects, leading to broad and confusing variations in results and misinterpretations of data. First, the so-called 'purified' diets of protein, carbohydrate, fats, and minerals were often impure and could, as Osborne and Mendel said of Röhmann's work, be contaminated with different vitamins. Second, the experimental animals varied from species to species in their need for certain vitamins; unlike humans, mice, rats, chickens, pigeons, and dogs can synthesize their own vitamin C and do not require it in the diet, but guinea pigs do require vitamin C. Third, different animals' ability to store certain vitamins in their tissues was not well understood, hence difficult to take into account. (Not until later did scientists become aware that the liver can store large but varying amounts of vitamin A, and this capacity was not calculated into interpretation of the results of experimental deprivation of dietary vitamin A [68].) And fourth, until large-scale breeding colonies were established, the individual animals purchased from suppliers could differ significantly from one another. The biochemist Casimir Funk (1884–1967), for example, reported in 1916 that 80% of the rats acquired for an ongoing experiment arrived at his laboratory with defects [69]. Such technical difficulties in performing experiments and interpreting results could impede progress and fuel rivalries and animosities.

Mendel articulated the problem caused by impure experimental diets in a 1914 lecture before the Harvey Society of New York: 'It is not unlikely – to speak conservatively – that there at least two 'determinants' in the nutrition of growth. One of these is furnished in our 'protein-free milk' which insures proper maintenance even in the absence of growth. . . another determinant is furnished by these natural fats (butterfat, cod liver oil, or egg fat). . . both are essential for growth when the body's store of them (if such there be) becomes depleted' [70]. As though to corroborate Mendel's cautionary note, McCollum and Davis began to realize in 1915 that the commercial lactose preparation they were providing as a 'purified' sugar in their basal diet was in fact impure [71]. An unidentified water-soluble factor was associated with the lactose [72]. Likewise, Mendel's own experiments, conducted with Osborne, came under question when it became apparent that the 'protein-free milk' they used was contaminated with other growth-supporting substances; the unknown contaminant turned

Fig. 5.4. Experimental rickets in dogs. From Mellanby, E. (1939). The experimental method in the conquest of disease. The thirteenth Stephen Paget Memorial Lecture. Fight Against Disease 27.

out to be thiamin, the B complex vitamin missing in patients with beriberi [73]. In fact, all four researchers had inadvertently tainted their rats' 'purified diets' with small amounts of thiamin.

While researchers in the United States doggedly searched for the causes of diseases, a British colleague turned his attention towards enigmatic remedies. Cod liver oil and butter had proven excellently but incomprehensibly effective in curing night xerophthalmia and rickets. But what did they contain that produced this salutary effect? At the request of the Medical Research Council, Edward Mellanby (1884–1955), Professor of Pharmacology at the University of Sheffield and former student of F. G. Hopkins, began to study rickets in dogs. Mellanby fed puppies a basal diet of milk (<200 ml/day), rice, oatmeal, and salt, or milk and bread. The addition to the basal diet of certain foods, such as cod liver oil, butter, or extra milk (500 ml/day), could prevent rickets in young puppies, whereas meat protein, casein, linseed oil, or yeast did not (fig. 5.4) [74]. Mellanby published the conclusion in 1918–1919 that rickets was a nutritional deficiency disease, and he attributed it to 'a diminished intake of an anti-rachitic factor which is either fat-soluble A, or has a somewhat similar distribution to fat-soluble A' [75]. His data did not exactly fit the idea that the factor, when missing, was what McCollum would term 'fat-soluble A'; Mellanby therefore qualified his statement about the distribution of this unknown factor in foods as possibly being *similar to* 'fat-soluble A.'

McCollum, meanwhile, was about to do an about face. Since biochemist Casimir Funk, like Mellanby working in England at the time, had suggested that deficiencies of 'vitamines' were the cause of beriberi, scurvy, and pellagra, McCollum prepared to study scurvy in his albino rats. After several experiments, McCollum made a startling declaration in 1917: scurvy was not a nutritional deficiency disease [76], despite rather overwhelming evidence to the contrary. McCollum and his research assistant, William Pitz, fed both rats and guinea pigs a diet of protein, sugar, butterfat (containing 'fat-soluble A') and wheat germ (containing 'water-soluble B'). The rats – which, unknown to McCollum do not require vitamin C – survived in perfect health, but the guinea pigs developed the typical symptoms of scurvy.

After testing additional diets and conducting pathological, germ-theory-based investigations, McCollum concluded that scurvy is not a nutritional deficiency at all; rather, it is caused by a bacterial infection brought on by extreme constipation. 'The significance of this interpretation is far reaching', he pronounced. 'It removes from the list one of the syndromes (scurvy) which has long been generally accepted as being due to a dietary deficiency.' McCollum continued to adhere to this idea for some time, perhaps because, if the rat could not be used to generalize about human dietary deficiencies, this might cast doubt on all the work he had done with 'fat-soluble A' and 'water-soluble B' [77].

In retrospect, McCollum's reversal might be construed as personally adversarial. While research was progressing quickly in the field of 'vitamines', personal disagreements and rivalries were beginning to surface. In 1917, at the invitation of Professor William Howell, McCollum moved from Wisconsin to Johns Hopkins University in Baltimore to head the Department of Chemical Hygiene at Johns Hopkins's newly founded School of Hygiene and Public Health. A cloud of accusations of ethical impropriety and professional misconduct accompanied McCollum's departure from Madison and was aired in print in the journal *Science* [78].

Upon McCollum's departure, all the research notebooks in the Wisconsin agricultural station disappeared. His former supervisor, Professor Hart, wrote a letter to *Science* titled 'Professional Courtesy' [79] that drew attention to the missing notebooks. Among them were those of Harry Steenbock, who was considered McCollum's 'ace student' and a 'brilliant' upcoming scientist [80]. Steenbock had worked with McCollum since 1915, 'on the understanding that opportunity would be given Steenbock to develop some independent problems which he could work out separately from McCollum, if he so desired' [81]. Before leaving Wisconsin, McCollum had reassured Dean Harry L. Russell that he would 'leave the records in such shape so that all Station material could be utilized'.

With McCollum now departing, Russell was worried about the future of the laboratory work, and he recorded his meeting with McCollum in his diary. Russell had cause to be concerned, as he wrote in his diary, 'Steenbock reported to me earlier in the day the conversation which he had had with McCollum in which McCollum was averse to giving him much of any information, said that from now on they were scientific rivals' [82].

Steenbock's research plan and data from his missing notebooks soon appeared in the *Journal of Biological Chemistry* – in a paper published by McCollum [83]. The paper was based on laboratory work carried out entirely by Steenbock and included wording taken verbatim from Steenbock's notebooks. But Steenbock name did not appear even as a co-author, nor had McCollum sought Steenbock's permission to publish his data.

McCollum responded to the ethical criticisms raised by the letter in *Science* with a dismissal. 'We do not feel that a reply to the charges contained in his statement is necessary, further than to say that the work referred to was *planned* (emphasis added) entirely by one of us (McCollum) and was carried out by Mr. Steenbock. . . (N)othing better can be done than to leave the public to judge for itself on the basis of the research records of all concerned as to the probable responsibility for the *planning* (emphasis added) of this work' [84]. Steenbock, in turn, pointed out that the Wisconsin Agricultural Experiment Station required that all scientific manuscripts be subject to review and approval by the director of the station, i.e. McCollum's onetime mentor, Edwin Hart. In violation of the policy, McCollum submitted two papers for publication without approval [85]. In a second letter to *Science*, again not responding to the specific allegations, McCollum gave a long explanation about how he alone built up the line of experimentation and the rat colony. He expressed the belief that his work had benefited large groups of malnourished people. He further claimed, 'during my stay at the University of Wisconsin nobody had anything to do with independent work with my rat colony. . . there is no property right in research or its results so long as it is incomplete and not protected by patent' [86].

Finally, McCollum took a parting shot at his one-time colleagues at Wisconsin. 'A few prefer to attempt to bring into disrepute some investigator who has opened up a new field of research when he has reached a point where much further work remains to be done. . . in the hope that they may thereby so discredit him that his work will be interfered with, with a view to making possible the reaping of a harvest of opportunity which his absence from the field would make possible'. A former staff member at Wisconsin recalled, '(T)his friction developed because Steenbock had become ambitious too and it was an impossible situation to have two forceful men in the same department. Especially when one of them, Steenbock, had occupied the position of subordinate before the antagonism developed' [87]. Steenbock, who had an impeccable reputation for honesty, wrote to Russell that he only wanted to set the record straight: 'I am not making capital out of this affair, but I do want to submit some facts and then quit' [88].

Given the tension that had grown up, McCollum's former colleagues at Wisconsin had been 'most pleasantly relieved by his departure' [89]. Despite their former collaborative relationship, Hart described McCollum as 'a poor operator on a team', tempering this criticism by noting that McCollum was 'a capable individualist' [90].

McCollum went beyond removing all the research notebooks, including those of other investigators' ongoing studies. As a final act of sabotage, he released all the albino rats from their cages in the animal colony at the Wisconsin Agricultural Experiment Station laboratories. Steenbock had to spend two full months trapping enough of the rats to restart the animal colony and resume his research program [91].

In New Haven, Mendel was certainly aware of the personal conflict that was brewing, since he had mentored both McCollum and Steenbock. Still collaborating with Osborne on their research, Mendel steered clear of the controversy, however. The work on which they were focused was bearing particularly valuable fruit: their 1918 studies identified the liver as the storage organ for vitamin A. To do so, they fed rats different organs and tissues from pigs, and thus found that liver was an extremely potent source of vitamin A [92]. At the time, scientists were beginning to appreciate the significance of the liver as the storage organ for vitamin A. The time it took for vitamin A deficiency to appear in their experimental animals depended upon how much vitamin A was stored in their livers.

During his first year in Baltimore, McCollum worked on a nutrition textbook and established another breeding colony of rats with which to continue his investigations. The book, published in 1918 and titled *The Newer Knowledge of Nutrition: the Use of Food for the Preservation of Vitality and Health*, included the author's criticism of the work of his colleague, Casimir Funk. McCollum asserted that his own studies had disproved the existence of some so-called deficiency diseases and that some of these diseases were more likely attributable to an improper proportion of proteins, carbohydrates, fats, and minerals:

> What has been said. . . regarding the special dietary properties of the different food-stuffs which go to make up the diet of civilized man, and the dietary habits of those classes of people who suffer from the diseases which have come to be recognized as being due to faulty diet, make it easy to see that there has become fixed in the minds of students of nutrition and of the reading public, an altogether extravagant idea regarding the importance of the substances which Funk gave the name 'vitamines'. Of the diseases which Funk considered due to lack of unidentified substances of this nature, namely beriberi, scurvy, pellagra, and rickets, but one, beriberi, has been shown to be due to this cause. . . Pellagra, scurvy and rickets do not belong in the same category with beriberi, and there do not exist 'curative' substances of unknown nature for these diseases. The individual is predisposed to the development of these syndromes by faulty diet, but the faults have been shown by the biological method for the analysis of the individual food-stuffs or their mixtures, to reside in maladjustments, and unsatisfactory quantitative relationships among the now well-recognized constituents of the normal diet. They are to be sought in the quality and quantity of the protein, the character and amount of the inorganic constituents. . . [93].

In his book (which sold fourteen thousand copies in the first three years), McCollum advocated greater consumption of what he termed 'protective foods' – milk, eggs, and leafy vegetables. The 'newer knowledge of nutrition' of the title and the basis of his recommendations came mostly from McCollum's nearly three thousand feeding experiments with his rat colony. But the book was not kindly received by some of his colleagues working in the field. In it, McCollum mentioned his own name nearly seventy times and, feigning a kind of remote objectivity, referred to himself in the third person. Few of the dozens of major investigators contributing to the 'newer knowledge' received more than passing mention, including F.G. Hopkins, Osborne and Mendel, Stepp, and his own one-time collaborator, Steenbock (see also textbox 5–4).

Even as McCollum dispensed from on high his advice on healthful eating, pellagra began to appear in certain populations in the southern part of the United States. Known previously in southern Europe and familiar in medical circles everywhere for its symptoms (the '4Ds', dermatitis, diarrhea, dementia, and death), pellagra made an unwelcome arrival in the United States and an aggressive advance in the South. With nearly sixteen thousand cases reported from eight states between 1907 and 1911, pellagra reached epidemic proportions, giving rise to considerable anxiety. The US Surgeon General asked Joseph Goldberger (1874–1929), an infectious disease specialist then working for the National Health Service, to investigate the causation of pellagra and allocated nearly one-fourth of his budget to the problem.

At the time, pellagra was widely believed to be an infectious disease, but it did not fit the picture of a contagious disease, and Casimir Funk said as much. Working counter to the common belief and following Funk's conviction, Goldberger conducted epidemiological investigations showing that pellagra was associated with the diet that predominated among the southern poor, namely, salted pork fat, corn bread, and molasses. Observing groups of employees and inmates incarcerated in prisons, asylums, and orphanages where pellagra was present, Goldberger found that even workers whose jobs brought them into close contact with pellagra victims – including himself – never contracted the disease [94]. He went on to demonstrate that pellagra could be eliminated by providing milk, eggs, fresh meat, beans, peas, and oatmeal in the diet [95]. At the same time, other studies were coming to opposite conclusions. The Thompson-McFadden Pellagra Commission, a research group funded by two philanthropists, conducted its own investigations in mill villages in the cotton belt in the south and concluded that pellagra was indeed an infectious disease [96].

Using more rigorous dietary methods than the commission's, Goldberger conducted further epidemiological investigations in the mill villages. Unlike the commission, he showed that households that consumed more lean meat, milk, butter, cheese, and eggs had lower risk of pellagra [97]. Furthermore, to test whether pellagra was an infectious disease, Goldberger and other volunteers (including his wife) went so far as to inject themselves with blood, nasal secretions, ground-up skin lesions, urine, and feces from pellagra patients: none of the volunteers developed the disease [98]. Goldberger's conclusion: 'On the whole. . . the trend of available evidence strongly suggests that pellagra will prove to be a "deficiency" disease very closely related to beriberi' [99]. Sound though his evidence was, however, Goldberger's conclusions did not meet with total acceptance.

On April 25, 1919, McCollum read a paper before the American Philosophical Society in which he criticized Goldberger's work, finding fault with most of the latter's studies and the very idea that pellagra could be caused by a lack of an unknown dietary substance. In the face of overwhelming evidence from epidemiological observations and intervention studies in humans, McCollum still would not accept that pellagra could be attributable to the lack of an unknown dietary substance. McCollum sided with the Thompson-MacFadden commission's view and declared that, 'pellagra is caused by an infectious agent' [100]. He took the occasion to reassert his exclusive confidence in animal experimentation and warned against drawing conclusions from dietary experiments with human adults. His position: '(W)e must be guided in human nutrition by the results of animal experimentation, in which the conditions can be made sufficiently rigid to bring into stronger contrast the faults of certain types of diets as contrasted with others'. McCollum and his colleagues had failed to produce pellagra in rats; hence, pellagra could not be a dietary deficiency disease.

Despite the self-assurance implied by this proclamation and others, McCollum grew increasingly nervous about his reputation. The scientific grapevine was aquiver with comment on his misbehavior while leaving Wisconsin. The assessment of his colleagues not only about his scientific accomplishments abut also about his character could potentially make the difference between recognition and high regard, and failure. The selection process for a Nobel Prize involved collecting the opinions of eminent scientists in a candidate's field – in McCollum's case, it could mean the opinion of highly respected Thomas Osborne in New Haven.

In Spring 1919, McCollum wrote with some urgency to Osborne's close collaborator and his former mentor, Lafayette Mendel [101]:

Dear Doctor Mendel,
I want to thank you for the copy of the collected reprints which you sent me recently. I prize this set which has been accumulating year by year very much. I shall keep you supplied with a set of my papers from time to time.

There is a matter which has caused me much concern lately and about which I have finally decided to ask your advice. It has come to me from several sources that Dr. Osborne has said some very hard things about me, and I am at a loss to know why he should do so. To be sure we have disagreed with you on certain matters in our research, but I have confined my criticism entirely to specific points of a highly technical nature, and, I believe, have had a basis for such criticism in experimental work in which I have firmly believed. I have never made uncomplimentary remarks about either of you, nor have I felt inclined to do so. I have always valued the friendship of both of you and have repeatedly mentioned your research in my public lectures and never in any instance have I verbally criticized it in any way. It has seemed to me that progress in science must be our goal and that in our technical articles we must frankly seek to place before our readers what we conscientiously believe to be the facts and their correct interpretation. This has been possible for me without in any way letting personal feeling enter into my attitude.

The latest one to report this unpleasant attitude on the part of Dr. Osborne was Dr. Howell. He did not tell me just what he said but indicated that it was so serious that he felt that I should take the matter up with Dr. Osborne and come to some sort of an understanding. Naturally I cannot rest until I have made an effort to set some matters right. Will you be so kind as to tell me frankly what you know about the situation, and advise me as to what I should do. I want to do the right thing by all, but it is a most serious thing for me to have accusations such as I have been told of made by a man who is held everywhere in such high regard as is Dr. Osborne.
With kindest regards,
Sincerely yours,
E. V. McCollum

Mendel was known for his kindness and had a reputation as a keen judge of character [102]. When new academic positions opened up across the country, Mendel was often consulted regarding his opinion of the possible candidates. Mendel was concerned for the success of all of his students, and his own students and fellows were highly sought after by other universities. He responded to McCollum a few days later [103]:

My Dear McCollum

I have your letter of May 20th in which you ask my advice regarding an unpleasant situation that seems to have arisen. At the outset I wish to emphasize what I feel confident is clear to you, namely, that I myself dislike exceedingly to be concerned in any way with personal differences, fancied or real, that are likely to impair the friendships which I value. My preference is to stand aloof from controversial discussions. However, in the present instance I presume that you have approached me in a spirit of personal friendship which I cannot disregard. Respecting of my own attitude you are perhaps aware that I have urged your advancement or appointment to scientific posts on more than one occasion. I shall therefore reply frankly, in the belief that you will read this unwilling expression of opinion in the light of correspondence between good friends and with a recognition of its confidential character.

You write: 'Will you be so kind as to tell me frankly what you know about the situation, and advise me as to what to do'.

I myself have never heard Osborne say anything detrimental to your character; in fact, outside of our own laboratory I cannot recall having heard him discuss you otherwise than incidentally. This does not mean that he and we have not often discussed your work, your results, methods and mode of presentation. On the contrary your papers have formed a frequent topic of conversation, as might be expected. Sometimes one or several of us have taken vigorous exception to some aspect of your work; often we have sought to reconcile apparent discrepancies between our results, and not infrequently your papers or certain groups of experiments or certain generalizations have called forth undiluted praise.

I believe, however, that Osborne has been frankly outspoken with respect to one aspect of your attitude towards the subject of nutrition – and in this he has not stood alone. From year to year your publications have revealed what seems to be a growing studied indifference to the contributions of other persons to the development of the science. The climax was reached in your recent book which (at least, so it intimated) seemingly makes you alone responsible for the newer progress in nutrition. It has been regarded by some as an ungenerous presentation that is oblivious to much that cannot be attached to your own valuable work. I am writing here in an impersonal way, for I have never reviewed or discussed your book in public. Hence you may know that I am not the author of any published notices thereof. That in the direction indicated your book has created an unfortunate impression elsewhere is apparent from reviews which doubtless you have seen. Under the circumstances I ought perhaps to tell you in confidence that comparable criticism have been repeated to me as coming from others, including even your own colleagues and former students. The upshot of this is that the impression of a somewhat self-centered viewpoint of the physiology of nutrition on your part is not confined to New Haven.

The tone of some of your published experimental criticisms, which have doubtless been worded strictly within the limits of journalist correctness, has sometimes disclosed (or has been interpreted to exhibit) a desire to belittle others in unnecessary ways. Science grows in part by correcting and supplementing earlier work, not primarily by disparaging it. I must confess that some of your pronouncements seem extremely cocksure, even to me.

I do not know what Dr. Howell has said to you. If I were to advise you, dear McCollum, as you request in your letter, it would be in this spirit: Your work has aroused widespread interest and approval. Keep it up vigorously. But if you are convinced that the judgment of various other workers and students to which I have alluded may have some justification, however slight, bear it in mind in the future, and be as tolerant, as generous and as sympathetic in your presentations as you very rightfully expect others to be of your own contributions. I am certain that you should gain true friends thereby. Imbued with this spirit, I should regard past incidents as closed.

With assurance of my personal regard, I am.

Lafayette B. Mendel

McCollum appears to have given some consideration to Mendel's advice. In subsequent editions of his book, he made more frequent mention of other investigators, though often in the context of his harsh criticism of their experiments. But McCollum continued to avoid citing important work of former colleagues at Wisconsin (textbox 5–5), prompting Hart to write to Mendel about these omissions: 'Evidently the University of Wisconsin is *persona non grata* among some of our explorers in biological chemistry. . . it is all very interesting in depicting the character of men, even scientists' [104].

Textbox 5–5. The lone wolf who never lost his bite

Elmer McCollum pursued his research at Johns Hopkins University in a style that his former dean Allen W. Freeman characterized as 'a lone wolf who liked to have assistants, usually women, rather than co-workers' [105]. McCollum felt that teaching interfered with his research. He gave only one lecture a week and complained that, 'A man in research has to have time for thought and reflection' [106]. His aversion to teaching translated into a sparse generation of graduate students and post docs to carry on nutritional research [107].

In 1957, McCollum published *A History of Nutrition: the Sequence of Ideas in Nutrition Investigations*. In it, he reiterated his claim to have 'discovered' both vitamins A and D [108]. Despite its appearance of comprehensiveness, the book has some obvious omissions, particularly if the work involved a Nobel Prize given to his colleagues. Missing are references to Paul Karrer's descriptions of the chemical structures of carotene and vitamin A, the important description of the chemical structure of vitamin D by the Nobel laureate Adolf Windaus, and the chemical description of vitamin C by Norman Haworth. He dismissed or disparaged colleagues' work, including the pivotal experiment of F.G. Hopkins of 1912, writing, 'Hopkin's [sic] experiments did not advance knowledge beyond what had been [already] proven. . .' [109]. He belittled Harry Steenbock's 1924 discovery that ultraviolet irradiation of foods generated vitamin D, noting – incorrectly – that Steenbock had merely 'confirmed' other investigators' findings [110].

Lafayette Mendel's Far-Flung Progeny and His Legacy

Contrary to Edwin Hart's quip that Wisconsin had been made an unwanted presence in biochemistry, the work in progress in the lab of his junior colleague Harry Steenbock was putting Madison front and center in the field. In 1918, on the basis of rat experiments, Steenbock published the *Journal of Biological Chemistry* the first in a remarkable series of papers titled 'Fat Soluble Vitamine'. Early work on 'fat-soluble A' dealt mostly with the vitamin A that occurs in foods from animal sources, that is, *preformed* vitamin A [111]. The primary sources of preformed vitamin A are milk, butter, cheese, liver, and cod liver oil. In fact, however, fat-soluble vitamin A also occurs

Table 5.2. Vitamin A content of foods is linked to yellow color

Vitamin A potency of foods	
Active (yellow)	Inactive or low (white)
butter	casein
egg yolk	egg white
cod liver oil	lard
yellow corn	white corn
carrot	parsnip
sweet potato	white potato
red palm oil	cottonseed oil
apricot	apple

in three related carotenoid molecules (alpha-carotene, beta-carotene, and beta-cryptoxanthin, known as provitamin A carotenoids) that are present in dark green leafy vegetables, and orange and yellow fruits and vegetables such as carrots, mango, and papaya. In April, at a meeting of the American Society of Biological Chemists in Baltimore, Steenbock announced the novel and important idea that the vitamin A levels in foods were related to the amount of yellow pigment present. In a series of papers on carotenoids, Steenbock showed that carrots and yellow sweet potatoes are sources of vitamin A, whereas white tubers such as rutabagas, parsnips, and white potatoes contain no vitamin A, nor do sugar beets (table 5.2) [112]. Yellow maize, but not white maize, contains vitamin A [113]. Dark leafy greens are rich sources of the vitamin [114]. (That the predominant blue-green of chlorophyll conceals yellow pigments was already known [115]. The yellow pigment of these leafy greens becomes evident in the fall, when the chlorophyll fades and no longer masks the yellow.) In fact, the yellow and orange hue of foods was found to be a good approximate indicator of vitamin A potency. For example, red palm oil, which is intensely deep red, is rich in vitamin A [116]. Oranges, in contrast, have only modest amounts [117].

A noticeable inconsistency arose from the observations of Steenbock and others: while yellow pigment is associated with high-potency vitamin A, cod liver oil, which is colorless, is also an extremely potent source of vitamin A. So the obvious question: If color does not offer a reliable way to detect the presence of vitamin A and its potency, what is a good index? To find an answer to that question, scientists had to determine exactly what constitutes the enigmatic 'fat-soluble A'.

In 1926, a color test was developed that allowed scientists to measure the amount of vitamin A present in solutions [118]. Antimony trichloride gave a deep blue color reaction that could be measured consistently. Soon the antimony trichloride color reaction and other methods, such as passing different wavelengths of light through solutions, were applied to contrast the differences between carotene and vitamin A (table 5.3). The controversy over the relationship between carotene and vitamin A

Table 5.3. Differences in carotene and vitamin A to the biochemist

Carotene	Vitamin A
deep red color	colorless
synthesized in plants	formed in the animal body from dietary carotene
ultraviolet light	ultraviolet light
absorption bands: 492, 462, 437, 348, 280 nm	absorption band 328 nm
reaction with antimony trichloride:	reaction with antimony trichloride:
dull blue	vivid blue

continued until 1928, when Elisabeth, Baroness of Ugglas, and her husband, Hans von Euler-Chelpin, at Stockholm University showed that carotene could cure vitamin A deficiency in rats [119]. In 1929, Thomas Moore at Cambridge University demonstrated that carotene could be converted to vitamin A. Moore fed crystalline carotene to rats and showed that the vitamin A concentration increased dramatically in the liver [120].

Another unsolved mystery still remained, however. Mellanby had noted in his rickets experiments with dogs that the data suggested that the unknown causal dietary factor had a distribution in foods that was somewhat similar to but not identical to that of vitamin A. 'Fat-soluble A' seemed resistant to heat alone at temperatures up to 120°C, but it was not indestructible. F. G. Hopkins had shown in 1920 that a combination of aeration (i.e. bubbling air through a liquid) and heat could oxidize 'fat-soluble A' and thus destroy its biological activity [121]. Two years later, McCollum and colleagues found when working with rats that if cod liver oil were aerated and heated, it no longer cured xerophthalmia but promoted calcium deposition in bone in rats with rickets. They thus showed that while cod liver oil consisted of 'fat-soluble A', which cures xerophthalmia, the oil also contains another fat-soluble substance, which plays a role in bone growth [122]. The latter fat-soluble substance was initially called the 'anti-rachitic factor', or simply 'X', and it eventually became known as vitamin D [123].

The distinction between vitamins A and D became more apparent with further rickets investigations. Sunlight deprivation was thought to play a role in rickets, and in 1919 a German pediatrician had shown that ultraviolet light cured children of rickets [124]. In that same year, Mendel received a letter from the British nutritionist Harriette Chick (1875–1977), who was the first woman appointed to London's Lister Institute of Preventive Medicine. Chick worked as assistant to the director, Charles Martin, and established a group to study beriberi and scurvy. She served as Secretary to the Accessory Food Factors (Vitamines) Committee, which had been established in 1918 with F.G. Hopkins as chair. At the time this was the world's only formal group devoted entirely to vitamin research. In her letter, Chick informed Mendel, 'I have just arrived in Austria on behalf of the Medical Research Committee to study the

Fig. 5.5. Rickets among children in Vienna. Two children with rickets with a healthy child in the center. From Medical Research Council (1932) Vitamins: a survey of present knowledge. London, His Majesty's Stationery Office.

"deficiency" diseases here and if possible to help in treatment, especially in cases of children. Rickets is terribly bad and scurvy seems to occur every winter. Of course shortage of milk is the real difficulty. . . and their diet after leaving the breast is very short and in some cases appears to be devoid of fat soluble A. It appears that seventy to ninety percent of the children between one and five years have rickets and many have terribly severe cases' [125].

The conditions Chick described were the result of cataclysmic upheaval in Europe. By the end of World War I and the fall of the Austro-Hungarian Empire, when Hungary had ceased to export food to Austria, hunger in Austria was reaching a critical level. Food for Vienna's population of 2.3 million was minimal [126]. The job of Chick's committee was to provide expertise on the nutritional needs of war-affected civilians. At first, Chick and her colleagues encountered skepticism at the University of Vienna's children's hospital. The director, Clemens von Pirquet, later wrote that he had 'little expectation that it would lead to results of much practical value. . . I was of the opinion that a vitamin deficiency in our ordinary diet was a very exceptional occurrence. . . (W)ith regard to the etiology of rickets I held the view that it was an infectious disease, widely prevalent in this part of Europe. . .' [127]. Deeply troubling reports of rickets among Vienna's children were reaching the world (fig. 5.5). Infants with corneal ulceration, keratomalacia, and blindness, too, were common [128]. Relief efforts began in earnest.

In 1920, Pirquet also wrote to Mendel on the progress of the nutritional investigations, which had given rise to a large and productive research enterprise. 'Besides the 300,000 children we are now feeding,' Pirquet wrote, there were '100 university professors who deserve feeding at least as much as the children' [129]. Pirquet reserved one-fifth of the hospital for rickets research. Under the auspices of the Accessory Food Factors Committee, Chick and her colleagues conducted a series of trials. They found that cod liver oil prevented or cured the children's rickets, and that the same result could be achieved by exposing the children to a mercury vapor lamp, which emits ultraviolet light. Whether from a mercury lamp or direct sunlight, ultraviolet light generates vitamin D in skin that is exposed to the light. The committee's conclusion: rickets was caused both by lack of certain foods and by inadequate exposure to sunlight [130]. (One of Chick's later achievements was development of standards for vitamin requirements; textbox 5–6).

> **Textbox 5–6.** Finally, vitamins find a place on the international map
>
> In June 1931, the first International Conference on Vitamin Standards was held in London under the auspices of the League of Nations (forerunner of the United Nations). Improving health on a multinational scale was one of the tasks the league undertook, and, by holding the vitamin standards conference, the league was making a tacit statement that the field of vitamins had reached maturity and that vitamins and physical wellbeing were fundamentally linked.
>
> That dual recognition translated into a call for measurement units and a set of standards as to how many units of each known vitamin were essential to maintaining good human health. Thus, the conference was charged with finding standards and recommending measurements units for 'fat soluble A, antirachitic vitamin D, antineuritic vitamin B, and antiscorbutic vitamin C' [131]. Edward Mellanby chaired the conference, which convened scientists from Great Britain, Sweden, Denmark, the Netherlands, Norway, France, Germany, and the United States. Many of the investigators who had figured prominently in the vitamin research of the previous two decades including were present, including Elmer McCollum, Harry Steenbock, Harriette Chick, Jack Drummond, and two Nobel laureates, Adolf Windaus and Hans von Euler-Chelpin (see fig. 5.6).
>
> The committee laid the fundamental groundwork for work that eventually led to the determination of the required daily allowances of the different vitamins. The most important initial step was to agree on how the vitamins were to be measured and to exchange reference materials (solutions or compounds with known amounts of vitamins) between the different laboratories around the world. This process was something akin to agreeing on a common currency to be used by all groups.

During this period of fruitful vitamin research, investigators throughout the western world turned to Mendel for advice – Hopkins in Cambridge, England; Hart in Wisconsin; McCollum in Baltimore; Chick in London; Pirquet in Vienna, and many others. In his thoughtful and generous style, Mendel advanced the research by

Fig. 5.6. First meeting of the League of Nations Conference on Vitamin Standards, London, June 17–20, 1931. Front row, left to right: Louis S. Fridericia, Katherine Coward, Adolf Windaus, Elmer McCollum, Edward Mellanby, Hans von Euler-Chelpin, Edvard Poulsson, Harriette Chick, Harry Steenbock. Second row, left to right: unidentified, Jack Drummond, unidentified, unidentified, Lucie Randoin, Wallace Aykroyd, Carl Scheunert, Atherton Seidell, Barend P. C. Jansen, unidentified. Photo courtesy of the Wellcome Library.

inspiring other, younger scientists. He characterized the worst and the best research environment in a letter to a colleague and the University of Chicago. '(I)t means a lot to an institution to have a considerable number of young men who are growing up to their possibilities. One reason why we have had few such conditions is that our heads of departments are as a rule too self-centered and autocratic. The ideal 'chief' is a good promoter of talent in men' [132]. Acting on his convictions, Mendel gave his students opportunities to meet and be inspired by leaders in the field. He and Hopkins had a long and warm friendship, which is reflected in the notebook of a colleague: 'When Dr. F. Gowland Hopkins of Cambridge University visited the old [Yale] laboratory. . . Dr. Mendel told his students that Dr. Hopkins would be there and that they might have an opportunity to meet him if they would 'hang around'. Needless to say everyone hurried through his work to be ready to meet the great British pioneer in nutrition. One student, hurrying down the circular stairway with a tray full of bottles and glassware, stumbled and fell with a great crash just as Mendel and Hopkins came up the winding stairs underneath. Dr. Hopkins remarked jovially, 'I see you have designs on my life', immediately dissipating the embarrassment of the very red-faced student' [133].

In the spring of 1924, Mendel received a letter from Wisconsin with the news that Steenbock had discovered that the vitamin D concentration of certain foods could be increased by irradiating the foods with ultraviolet light. 'I am writing you this', Steenbock began, 'on the assumption that as my former instructor you are personally interested in my professional welfare and I accordingly invite your criticism' [134]. Mendel replied that his onetime protégé's as yet unpublished work:

. . . (has) awakened my ardent enthusiasm and filled me with pleasure. I congratulate and compliment you on the prospect of a tremendously important discovery. Furthermore, I appreciate very much the cordial spirit that you have shown in taking me into your confidence, so to speak, in connection with what you have been doing. I hope that you will clinch the essential finding regarding the effect of the radiation on non-potent fats so that there can be no question of any accidental error in reaching your conclusion; and then you should make sure that you are not deprived of the credit of the discovery to which you are entitled in connection with this investigation [135].

On September 5, 1924, Steenbock published a paper in *Science* that announced his findings [136].

With rickets highly prevalent in many parts of the world, Steenbock's process for increasing the vitamin D content of foods could be widely and lucratively applied in the food industry. Steenbock recognized the commercial implications and arranged to meet with the dean and president of the University of Wisconsin to urge them to seize the economic advantage and seek the patents for his process. Being either frugal or shortsighted, the university refused to pay for the patent applications, so Steenbock took the matter into his own hands and, for USD 660, hired an attorney to secure the patents. The Quaker Oats Company offered Steenbock USD 1 million for rights to the patents. But Steenbock believed that his professor's salary was sufficient for himself and that his university and scientific research should be the beneficiaries [137].

At Steenbock's insistence, the Wisconsin Alumni Research Foundation was eventually incorporated. Steenbock spent ten dollars to assign his patents to the foundation and requested no share of the royalties; his patents eventually earned the foundation some USD 14 million. Resolutely modest and unassuming, Steenbock returned to his lab, working with Hart and other colleagues, to continue a highly productive career. He made seminal scientific contributions, including further new insights into vitamins D and A, nutritional anemia, and vitamin E [137], while his vitamin D discoveries earned millions for the University of Wisconsin.

Not until 1931 was the chemical structure of vitamin A finally described. This was accomplished by Paul Karrer (1889–1971), an organic chemist in Zurich, who, the year before, had deduced the correct chemical structure of beta-carotene (Karrer was later awarded a Nobel Prize in Chemistry in 1937 for discovering the structures of vitamin A and beta-carotene) [138]. Six years later, Harry Holmes and Ruth Corbet at Oberlin College had crystallized vitamin A [139]. A full decade had passed by the time Otto Isler and colleague chemists at the Swiss pharmaceutical company Hoffmann-La Roche synthesized vitamin A [140]. The crystallization and synthesis were critical to understanding vitamin A, and these final steps paved the way for the production of pure preparations of vitamin A that could be used for vitamin preparations and the fortification of foods for the public.

The relationship between vitamin A deficiency and night blindness was finally solved definitively through the work of George Wald (1906–1997), a biochemist at Harvard University. Before working at Harvard, Wald had spent time in the biochemistry laboratory of Otto Meyerhof in Heidelberg. Wald had posited that rhodopsin, the 'visual purple' in the retina (see Chapter One), was related to the carotenoids; he based this theory on the absorption spectrum of rhodopsin. (When lights of different wavelengths pass through a compound in solution, the pattern of the light absorption yields information that can serve in deducing the structure of the compound.) The opportunity to test this supposition presented itself in Heidelberg. On a summer day when most of the lab personnel were on holiday, a shipment of three hundred frogs arrived, but no one seemed to be there to experiment with them. A lab assistant who, like Wald, had stayed behind to work was about to release the frogs when Wald stopped him [141]. With this windfall of study subjects, Wald was able to examine the frogs' retinas under diverse light conditions. In solutions of the visual purple, he found high concentrations of vitamin A [142].

Wald detected a novel carotenoid that was yellow in color, which he termed retinene. Retinine, he found, was present in retinas from animals with dark-adapted eyes. Conversely, he found high concentrations of vitamin A in retinas that had been exposed to light. These observations led Wald to proposed that the retina undergoes a visual cycle in which light causes the purple of rhodopsin to turn bright orange, which, in turn, fades to a yellow color ('visual yellow') consisting of retinene plus the protein opsin [143]. The visual yellow then loses color, becoming 'visual white,' which is rich in vitamin A. He proposed that rhodopsin is regenerated from vitamin A plus opsin or retinene plus opsin [144].

These observations by Wald brought to completion the long line of investigation that began with early observations relating night blindness and lack of vitamin A [145]. The characterization of vitamin A, from the hints of its existence shown by Magendie in 1816 to its isolation, description of chemical structure, and eventual synthesis in 1947, was an excursion down a tortuous road that took nearly one hundred and thirty years to complete.

References

1   Richardson, A. S. (1916) What the expectant mother should eat. Pictorial Review 17, 37.

2   Osborne, T. B., Mendel, L. B. (1911) Feeding experiments with isolated food-substances. Washington, D.C., Carnegie Institute of Washington, Publication No. 156, Part I, 6.

3   Snow, J. (1855) On the mode of communication of cholera. 2nd ed. London, John Churchill.

4   Budd, W. (1849) Malignant cholera: Its mode of propagation and its prevention. London, John Churchill.

5   Koch, R. (1876) Die Aetiologie der Milzbrand-Krankheit, begründet auf die Entwicklungsgeschichte des Bacillus anthracis. Beiträge zur Biologie der Pflanzen, Bd. II, Heft 2, 277–310.

6   Laveran, C. L. A. (1881) Un nouveau parasite trouvé dans le sang de plusieurs malades atteints de fièvre palustre. Bulletins et mémoires de la Société Médicale des Hôpitaux de Paris 2 sér. 17, 158–164.

7   Koch, R. (1882) Die Aetiologie der Tuberculose. Berliner klinische Wochenschrift 19, 221–230.

8   Koch, R. (1884) Ueber die Cholerabakterien. Deutsche medizinische Wochenschrift 10, 725–728.

9   Hansen, G. H. A. (1874) Undersøgelser Angående Spedalskhedens Årsager. Norsk magazin for Laegervidenskaben 4, 1–88.

10 Klebs, T. A. E. (1883) Ueber Diphtherie. Verhandlungen des Deutschen Kongress für Innere Mededizin 2, 138–154.

11 Details of this complicated scheme have been summarized by Carpenter, K. (1994) Protein and energy: a study of changing ideas in nutrition. Cambridge, Cambridge University Press, pp. 48–52.

12 Liebig, J. von (1842) Der Tierchemie oder die organische Chemie in ihrer Anwendung auf Physiologie und Pathologie, Braunschweig.

13 Carpenter (1994), p. 52.

14 Lunin, N. (1881) Über die Bedeutung der anorganischen Salze für die Ernährung des Thieres. Zeitschrift für physiologische Chemie 5, 31–39.

15 Bunge, G. von (1887) Lehrbuch der physiologischen und pathologischen Chemie. In zwanzig Vorlesungen für Ärzte und Studirende. Leipzig, F. C. W. Vogel, p. 105. (Translation from Bunge, G. von (1902) Text-book of physiological and pathological chemistry. Second English Edition. Philadelphia, P. Blakiston's Son & Co., p. 89.)

16 Socin, C. A. (1891) In welcher Form wird das Eisen resorbirt? Zeitschrift für physiologische Chemie 15, 93–139.

17 Hopkins, F. G., Cole, S. W. (1901) A contribution to the chemistry of proteids. Part I. A preliminary study of a hitherto undescribed product of tryptic digestion. Journal of Physiology 27, 418–428.

18 Hopkins, F. G. (1949) Autobiography of Sir Frederick Gowland Hopkins. In Needham, J. (ed.) (1949) Hopkins & Biochemistry (1861–1947). Cambridge, W. Heffer and Sons, pp. 3–25.

19 Stephenson, M. (1949) Sir F. G. Hopkins' teaching and scientific influence. In Needham, J. (ed.) (1949) Hopkins & Biochemistry (1861–1947). Cambridge, W. Heffer and Sons, pp. 29–38.

20 Willcock, E. G., Hopkins, F. G. (1906) The importance of individual amino-acids in metabolism. Observations on the effect of adding tryptophane to a dietary in which zein is the sole nitrogenous constituent. Journal of Physiology 35, 88–102.

21 Hopkins, F. G. (1906) The analyst and the medical man. Analyst 31, 385–397.

22 Hopkins, F. G. (1912) Feeding experiments illustrating the importance of accessory factors in normal dietaries. Journal of Physiology 44, 425–460.

23 Eijkman, C. (1890) VI. Polyneuritis bij hoenderen. Geneeskundig tijdschrift voor Nederlandsch-Indië 30, 295–334.

24 Eijkman, C. (1896) Polyneuritis bij hoenders. Nieuwe bejdrage tot de Aetiologie der Ziekte. Geneeskundig tijdschrift voor Nederlandsch-Indië 36, 214–269.

25 Grijns, G. (1901) Over polyneuritis gallinarum. Geneeskundig tijdschrift voor Nederlandsch-Indië 41, 3–110.

26 McCollum, E. V., Kennedy, C. (1916) The dietary factors operating in the production of polyneuritis. Journal of Biological Chemistry 24, 491–502.

27 Jansen, B. C. P., Donath, W. F. (1926) Antineuritische vitamine. Chemische Weekblad 23, 1387–1409.

28 Williams, R. R. (1936) Structure of vitamin B1. Journal of the American Chemical Society 58, 1063–1064; Williams, R. R., Cline, J. K. (1936) Synthesis of vitamin B1. Journal of the American Chemical Society 58, 1504–1505.

29 Semba, R. D. (2012) The historical evolution of thought regarding multiple micronutrient nutrition. Journal of Nutrition 142, 143S–156S.

30 Hulshoff-Pol, D. J. (1902) Katjang idjo, un nouveau médicament contre le béribéri. Janus 7, 524–534, 570–581.

31 Fraser, H., Stanton, A. T. (1909) An inquiry concerning the etiology of beri-beri. Studies from the Institute for Medical Research, Federated Malay States, No. 10. Singapore: Kelly & Walsh.

32 Bland Sutton, J. (1888) Rickets in monkeys, lions, bears, and birds. Journal of Comparative Medicine and Surgery 10, 1–29.

33 Routh, C. H. F. (1879) Infant feeding and its influence on life or the causes and prevention of infant mortality. Third edition. New York, William Wood & Company, 1879, p. 238.

34 Fischer, L. (1902) Infant-feeding in its relationship to health and disease. Second edition. Philadelphia, F. A. Davis Company, p. 284.

35 Holst, A. (1907) Experimental studies relating to 'ship-beri-beri' and scurvy. I. Introduction. Journal of Hygiene 7, 619–633; Holst, A., Frölich, T. (1907) Experimental studies relating to ship-beri-beri and scurvy. II. On the etiology of scurvy. Journal of Hygiene 7, 634–671.

36 Carpenter, K. J. (1986) The history of scurvy and vitamin C. Cambridge, Cambridge University Press, p. 174.

37 Mori, M. (1896) [On hikan, a type of marasmus] Chūgai iji shinpōh 386, 6–10; Mori, M. (1896) [Pathogenesis of so-called hikan and a specific medicine for it] Chūgai iji shinpōh 386, 554; Mori, M. (1904) Über den sog. Hikan (Xerosis conjunctivae infantum ev. Keratomalacie). II. Mitteilung. Jahrbuch für Kinderheilkunde und physische Erziehung 59, 175–195.

38 Semba, R. D. (2000) Théophile Roussel and the elimination of pellagra from 19th century France. Nutrition 16, 231–233.

39 Roussel, T. (1866) Traité de la pellagre et des pseudopellagres. Paris, J. B. Ballière et fils.

40 Funk, C. (1912) The etiology of the deficiency diseases. Beri-beri, polyneuritis in birds, epidemic dropsy, scurvy, experimental scurvy in animals, infantile scurvy, ship beri-beri, pellagra. Journal of State Medicine 20, 341–368.

41 Drummond, J. C. (1920) The nomenclature of the so-called accessory food factors (vitamins). Biochemical Journal 14, 660.

42 Stepp, W. (1909) Versuche über Fütterung mit lipoidfreier Nahrung. Biochemisches Zeitschrift 22, 452–460.

43 Stepp, W. (1911) Experimentelle Untersuchungen über die Bedeutung der Lipoide für die Ernährung. Zeitschrift für Biologie 57, 135–170. For a detailed account of Stepp's experiments, see Wolf, G., Carpenter, K. J. (1997) Early research into the vitamins: the work of Wilhelm Stepp. Journal of Nutrition 127, 1125–1259.

44 Osborne, T. B. (1924) The vegetable proteins. Second edition. London, Longmans, Green and Co., p. 4.

45 MS1146. Series I. Box 2. Folder 21, 1935–1941. Unsigned manuscript dated July 16, 1941, titled: Lafayette B. Mendel, Ph.D. LL.D. Great teacher and research worker in the field of nutrition. Rough notes on incidents which might be used for dramatized [sic] in the program 'Listen America.'

46 Osborne and Mendel (1911), Part II, pp. 76, 79.

47 Osborne, T. B., Mendel, L. B. (1912) Feeding experiments with fat-free food mixtures. Journal of Biological Chemistry 12, 81–89.

48 McCollum, E. V. (1964) From Kansas farm boy to scientist: the autobiography of Elmer Verner McCollum. Lawrence, University of Kansas Press, p. 102.

49 McCollum (1964), p. 114.

50 McCollum, E. V. (1952) Early experiences with vitamin A – a retrospect. Nutrition Reviews 10, 161–163.

51 McCollum (1964), p. 122.

52 McCollum, E. V. (1909) Nuclein synthesis in the animal body. American Journal of Physiology 25, 120–141.

53 Hart, E. B., McCollum, E. V., Steenbock, H., Humphrey, G. C. (1911) Physiological effect on growth and reproduction of rations balanced from restricted sources. Wisconsin Agricultural Experiment Station Research Bulletin No. 17.

54 McCollum, E. V., Davis, M. (1913) The necessity of certain lipins in the diet during growth. Journal of Biological Chemistry 15, 167–175.

55 McCollum, E. V. (1960) From Hopkins to the present. In Galdston, I. (ed.) Human nutrition historic and scientific. Monograph III. New York, International Universities Press, Inc., pp. 111–142; McCollum (1964), p. 134.

56 Wolf & Carpenter (1997), p. 1258.

57 Nobelstiftelsen (1930) Les Prix Nobel en 1929. Stockholm, Imprimerie Royale. P. A. Norstedt & Söner, p. 47.

58 Pekelharing, C. A. (1905) Over onze kennis van de waarde der voedingsmiddelen uit chemische fabrieken. Nederlands tijdschrift voor geneeskunde 41, 111–124. This paper consists of a lecture given to the General Assembly of the Netherlands Society for the Advancement of the Medical Arts in Amsterdam, July 3, 1905, and it contains scant details and no actual data regarding the milk experiments conducted by Pekelharing; Hopkins, F. G. (1930) Earlier history of vitamin research. Nobel Lecture delivered on December 11, 1929. In Nobelstiftelsen. Les Prix Nobel en 1929, Stockholm, Imprimerie Royale. P. A. Norstedt & Söner, p. 5.

59 Nobelstiftelsen (1938) Le Prix Nobel en 1937. Stockholm, Imprimerie Royale, P. A. Nordstedt & Söner, p. 35. When the Nobel Prize was awarded to Hopkins and Eijkman for the discovery of the vitamins, an editorial in Time Magazine (November 11, 1929) noted: 'Many a US nutritionist declared last week, without carping at the Nobel award to Professors Hopkins and Eijkman, that, if a future Nobel Prize for vitamin research is made, it should go to Professor Elmer Verner McCollum, 50, head of the department of chemical hygiene at Johns Hopkins School of Hygiene & Public Health.'

60 Osborne, T. B., Mendel, L. B. (1913) The relationship of growth to the chemical constituents of the diet. Journal of Biological Chemistry 15, 311–326.

61 Knapp, P. (1909) Experimenteller Beitrag zur Ernährung von Ratten mit künstlicher Nahrung und zum Zusammenhang von Ernährungsstörungen mit Erkrankungen der Conjunctiva. Zeitschrift für experimentelle Pathologie und Therapie 5, 147–169.

62 Osborne, T. B., Mendel, L. B. (1914) The influence of butter-fat on growth. Journal of Biological Chemistry 15, 423–437; Osborne, T. B., Mendel, L. B. (1914) The influence of cod liver oil and some other fats on growth. Journal of Biological Chemistry 17, 401–408.

63 McCollum, E. V., Simmonds, N. (1917) A biological analysis of pellagra-producing diets. II. The minimum requirements of the two unidentified dietary factors for maintenance as contrasted with growth. Journal of Biological Chemistry 32, 181–193.

64 McCollum, E. V., Simmonds, N., Parsons, H. T. (1918) A biological analysis of pellagra-producing diets. III. The nature of the dietary deficiencies of a diet derived from peas, wheat flour, and cottenseed [sic] oil. Journal of Biological Chemistry 33, 411–423.

65 Röhmann, F. (1916) Über künstliche Ernährung und Vitamine. Berlin, Gebrüder Borntraeger, p. 142.

66 Röhmann, F. (1908) Ueber künstliche Ernährung von Mäusen. Allgemeine medizinische Central-Zeitung 77, 129.

67 Osborne, T. B., Mendel, L. B. (1917) The rôle of vitamines in the diet. Journal of Biological Chemistry 31, 149–163.

68 Chick, H. (1926) Sources of error in the technique employed for the biological assay of fat-soluble vitamins. Biochemical Journal 20, 119–130.

69 Funk, C., Macallum, A. B. (1916) Studies on growth. III. The comparative value of lard and butter fat in growth. Journal of Biological Chemistry 27, 51–62.

70 Mendel, L. B. (1915) Nutrition and growth. The Harvey Lectures, delivered under the auspices of the Harvey Society of New York 1914–1915. Philadelphia, J. B. Lippincott Company, pp. 101–131.

71 McCollum, E. V., Davis, M. (1915) Nutrition with purified food substances. Journal of Biological Chemistry 20, 641–658.

72 McCollum, E. V., Davis, M. (1915) The essential factors in the diet during growth. Journal of Biological Chemistry 23, 231–246.

73 Drummond, J. C. (1916) The growth of rats upon artificial diets containing lactose. Biochemical Journal 10, 89–102.

74 Mellanby, E. (1918) The part played by an 'accessory factor' in the production of experimental rickets. Proceedings of the Physiological Society, January 26, 1918, xi–xii; Mellanby, E. (1918) A further demonstration of the part played by accessory food factors in the aetiology of rickets. Proceedings of the Physiological Society, December 14, 1918, liii–liv.

75 Mellanby, E. (1919) An experimental investigation of rickets. Lancet 1, 407–412.

76 McCollum, E. V., Pitz, W. (1917) The 'vitamine' hypothesis and deficiency diseases. A study of experimental scurvy. Journal of Biological Chemistry 31, 229–253.

77 Carpenter (1986) p. 183.

78 Hart, E. B. (1918) Professional courtesy. Science 47, 220–221; MS1146. Accession 1998-M-099. Box 1, Folder 1, Hart, E. B. Letter to L. B. Mendel, November 16, 1917.

79 Hart (1918); Series No. 9/11/13–2, Box 1, Folder 11. Steenbock, H. Letter to H. L. Russell, March 12, 1918.

80 PP120, Box 2. Brannon, W. A. Letter to E. O. Keiles, February 15, 1953.

81 Series No. 9/1/1/22–1, Box No. 4. Russell diaries, 'Dr. McCollum's Work,' April 26, 1915.

82 Series No. 9/1/1/22–1, Box No. 4. Russell diaries, 'Completion of McCollum's Work,' June 11, 1917.

83 McCollum, E. V., Simmonds, N. (1917) A biological analysis of pellagra-producing diets. II. The minimum requirements of the two unidentified dietary factors for maintenance as contrasted with growth. Journal of Biological Chemistry 32, 181–193.

84 McCollum, E. V., Simmonds, N. (1918) Professional courtesy. Science 47, 241.

85 Steenbock, H. (1918) Professional courtesy. Science 47, 535–536. In recounting the episode of the two disputed papers in his autobiography, McCollum alleges: 'Unfortunately I failed to say that they were published with permission of the Director, as was customary in experiment station bulletins,' [McCollum (1967), p. 150]. On the contrary, Russell's confidential notes of his meeting with McCollum on June 11, 1917 taken in his 'black book' diaries shows that he did not give permission for McCollum to publish the papers (Russell kept meticulous notes of his meetings with various faculty members in his diaries). Russell later sent a telegram to Steenbock on April 3, 1918 that reiterated the lack of authorization: 'McCollum agreed to submit results of Wisconsin work to me prior to publication did not give him permission to publish biological chemistry papers' [Series No. 9/11/13–2. Box 1. Folder 10. Russell, H. L. Telegram to H. Steenbock, April 3, 1918]. In addition, Hart took the extraordinary step of writing to Mendel, editor of the Journal of Biological Chemistry, to inform Mendel that McCollum specifically violated University of Wisconsin policy and published the papers without the required Station review or approval [MS1146. Accession 1998-M-099. Box 1, Folder 1, Hart, E. B. Letter to L. B. Mendel, November 16, 1917].

86 McCollum, E. V. (1918) Professional courtesy. Science 47, 536–538.

87 PP120, Box 2. Brannon, W. A. Letter to E. O. Keiles, February 15, 1953.

88 Series No. 9/11/13–2, Box 1, Folder 11. Steenbock, H. Letter to H. L. Russell, March 12, 1918.

89 MS1146. Accession 1998-M-099. Box 1, Folder 1, Hart, E. B. Letter to L. B. Mendel, November 16, 1917.

90 PP120, Box 2. Hart, E. B. Letter to E. O. Keiles, June 20, 1952.

91 DeLuca, H. E-mail communication to the author, June 22, 2009. Professor Hector DeLuca did his doctoral work under the supervision of Harry Steenbock from 1951–1954 and heard the account of McCollum's departure from the University of Wisconsin directly from Steenbock, who he says was 'extremely honest.' The episode is also consistent with the published research record. Steenbock was quick to publish his data, but he was only able to submit his first paper in the 'Fat-Soluble Vitamin' series on July 1918, nearly one year after the reported disruption of their research with albino rats.

92 Osborne, T. B., Mendel, L. B. (1918) Nutritive factors in animal tissues. II. Journal of Biological Chemistry 34, 17–27.

93 McCollum, E. V. (1918) The newer knowledge of nutrition: the use of food for the preservation of vitality and health. New York, Macmillan Company, pp. 113–114.

94 Goldberger, J. (1914) The etiology of pellagra. The significance of certain epidemiological observations with respect thereto. Public Health Reports 29, 1683–1686. Further details of Goldberger's investigations are summarized in Semba, R. D. (2007) The impact of improved nutrition. In Ward, J. W., Warren, C. (eds) Silent victories: the history and practice of public health in twentieth-century America. New York, Oxford University Press, pp. 163–192.

95 Goldberger, J., Waring, C. H., Willets, D. G. (1915) The prevention of pellagra. A test of diet among institutional inmates. Public Health Reports 30, 3117–3131.

96 Siler, J. F., Garrison, P. E., MacNeal, W. J. (1914) A statistical study of the relation of pellagra to use of certain foods and to location of domicile in six selected industrial communities. Archives of Internal Medicine 14, 292–373.

97 Goldberger, J., Sydenstricker, E. (1918) A study of the diet of nonpellagrous and of pellagrous households in textile mill communities in South Carolina in 1916. Journal of the American Medical Association 71, 944–949.

98 Goldberger, J. (1916) The transmissibility of pellagra. Experimental attempts at transmission to the human subject. Public Health Reports 31, 3159–3173.

99 Goldberger, J. (1916) Pellagra: causation and a method of prevention. A summary of some of the recent studies of the United States Public Health Service. Journal of the American Medical Association 60, 471–476.

100 McCollum, E. V. (1919) The relation of the diet to pellagra. Proceedings of the American Philosophical Society 58, 41–54.

101 MS1146. Accession 1998-M-099, Box 1, Folder 1. McCollum, E. V. Letter to L. B. Mendel, May 20, 1919. Mendel showed McCollum's letter to Osborne, who subsequently wrote a letter to Howell, explaining that his comments regarding McCollum were made privately and in confidence to Howell and to the Academy: 'When McCollum's name was before the Academy I assumed that it was not only my right but my duty to express my opinion of his work and his publications. . .I had always supposed that everything said or done in considering and discussing candidates for the Academy were matters of confidence between the members. . .I do not feel called on to apologize for anything I ever said about McCollum.' Connecticut Agricultural Experiment Station Archives. Osborne, T. B., Letter to W. H. Howell, June 9, 1919.

102 MS1146. Accession 1998-M-099, Box 1, Folder 6. Macy, I. G. Letter to L. B. Mendel, February 3, 1933; Series No. 9/11/13–2. Box 1, Folder 6. Steenbock, H. Letter to E. B. Gross, March 6, 1924.

103 MS1146. Accession 1998-M-099, Box 1, Folder 1. Mendel, L. B. Letter to Elmer V. McCollum, May 26, 1919.

104 MS1146. Accession 1998-M-099, Box 1, Folder 2. Hart, E. B. Letter to L. B. Mendel, March 18, 1924.

105 PP120, Box 2. Letter from A. W. Freeman to E. O. Keiles, November 14, 1952.

106 PP120, Box 2. Letter from A. G. Hogan to E. O. Keiles, September 1, 1952.

107 This can be seen clearly in the scientific genealogy of the American Institute of Nutrition, see Darby, W. J., Jukes, T. H. (1992) Founders of nutritional science: biographical articles from the Journal of Nutrition volumes 5–120, 1932–1990. Bethesda, MD, American Institute of Nutrition, Vol. 2, foldout facing p. 1182.

108 McCollum (1957), pp. 217–218, 281. In addition to claims for the 'discovery' of vitamins A and D, claims were made elsewhere that McCollum also 'discovered' vitamin B. For example, see Herriott, R. M. (ed.) (1953) Symposium on nutrition: the physiological role of certain vitamins and trace elements. Baltimore, The Johns Hopkins Press, p. xv.

109 McCollum (1957), p. 211.

110 McCollum (1957), p. 284. Other conspicuous omissions from McCollum's book include the work on alcoholic fermentation by Arthur Harden and Hans von Euler-Chelpin for which they received the Nobel Prize in 1929. McCollum introduces a false reference to dispute the priority of Joseph Goldberger in his studies of pellagra (pp. 303–304). McCollum often distorted the sequence of investigation or omitted their findings, especially that of his former competitors, in A History of Nutrition.

111 Preformed vitamin A consists of animal sources of vitamin A such as that found in liver, cod liver oil, egg yolk, butter, and cheese, in contrast to pro-vitamin A carotenoids found in plant-source foods such as dark green leafy vegetables and orange and yellow fruit and vegetables.

112 Steenbock, H., Gross, E. G. (1919) Fat-soluble vitamine. II. The fat-soluble vitamine content of roots, together with some observations on their water-soluble vitamine content. Journal of Biological Chemistry 40, 501–531.

113 Steenbock, H., Boutwell, P. W. (1920) Fat-soluble vitamine. III. The comparative value of white and yellow maizes. Journal of Biological Chemistry 41, 81–96.

114 Steenbock, H., Gross, E. G. (1920) Fat-soluble vitamine. IV. The fat-soluble vitamine content of green plant tissues together with some observations on their water soluble vitamine content. Journal of Biological Chemistry 1920, 41, 149–162.

115 Moore, T. (1957) Vitamin A. Amsterdam, Elsevier Publishing Company.

116 Drummond, J. C., Coward, K. H. (1920) Researches on the fat-soluble accessory substance. V: The nutritive value of animal and vegetable oils and fats considered in relation to their colour. Biochemical Journal 14, 668–677. Palm oil is a rich source of 'fat-soluble accessory substance.'

117 Osborne, T. B., Mendel, L. B. (1921) Vitamin A in oranges. Proceedings of the Society for Experimental Biology and Medicine 19, 187–188.

118 Carr, F. H., Price, E. A. (1926) Colour reactions attributed to vitamin A. Biochemical Journal 20, 497–501.

119 von Euler, B., von Euler, H., Hellström, H. (1928) A-Vitaminwirkungen der Lipochrome. Biochemische Zeitschrift 203, 370–384.

120 Moore, T. (1929) The relation of carotin to vitamin A. Lancet 2, 380–381; Moore, T. (1930) Vitamin A and carotene. V. The absence of the liver oil vitamin A from carotene. VI. The conversion of carotene to vitamin A in vivo. Biochemical Journal 24, 692–702.

121 Hopkins, F. G. (1920) The effects of heat and aeration upon the fat-soluble vitamine. Biochemical Journal 14, 725–733.

122 McCollum, C. V., Simmonds, N., Becker, J. E., Shipley, P. G. (1922) Studies on experimental rickets. XXI. An experimental demonstration of the existence of a vitamin which promotes calcium deposition. Journal of Biological Chemistry 53, 293–312.

123 Blunt, K., Cowan, R. (1930) Ultraviolet light and vitamin D in nutrition. Chicago, University of Chicago Press.

124 Huldschinsky, K. (1919) Heilung von Rachitis durch künstliche Höhensonne. Deutsche medizinische Wochenschrift 45, 712–713; Huldschinsky, K. (1919) Die Behandlung der Rachitis durch Ultraviolettbestrahlung. Dargestellt an 24 Fällen. Zeitschrift für orthopädische Chirurgie, einschliesslich der Heilgymnastik und Massage 39, 426–451.

125 MS1146, Series 1, Box 1, Folder 10. Chick, H. Letter to L. B. Mendel, September 28, 1919.

126 Dalyell, E. J., Chick, H. (1921) Hunger-osteomalacia in Vienna, 1920. Its relation to diet. Lancet 2, 842–849; Williams, H. (1918) Vienna facing cold and famine. The New York Times December 11, 1918.

127 Pirquet, C. [von] (1923) Preface. In [Chick, H.] Studies of Rickets in Vienna 1919–22 (Report to the Accessory Food Factors Committee appointed jointly by the Medical Research Council and the Lister Institute). London, His Majesty's Stationery Office, pp. 5–6.

128 Chick, H., Hume, M., Macfarlane, M. (1971) War on disease: a history of the Lister Institute. London, André Deutsch, pp. 155–156.

129 MS1146, Series 1, Box 1, Folder 11. Pirquet, C. Letter to L. B. Mendel, October 17, 1920.

130 Medical Research Council (1923) Studies of Rickets in Vienna 1919–22 (Report to the Accessory Food Factors Committee appointed jointly by the Medical Research Council and the Lister Institute). London, His Majesty's Stationery Office. For a summary of the work, see Carpenter, K. J. (2008) Harriette Chick and the problem of rickets. Journal of Nutrition 138, 827–832.

131 League of Nations Health Organisation. Permanent Commission on Biological Standardisation. (1931) Report of the Conference on Vitamin Standards (London, June 17th to 20th, 1931). Geneva, League of Nations, Document C. H. 1055 (I).

132 MS1146. Accession 1998-M-099, Box 1, Folder 2. Mendel, L. B. Letter to A. J. Carlson, March 10, 1924.

133 MS1146. Series I. Box 2. Folder 21, 1935–1941. Unsigned manuscript dated July 16, 1941. Hopkins visited Yale in the spring of 1921.

134 MS1146. Accession 1998-M-099, Box 1, Folder 2. Steenbock, H. Letter to L. B. Mendel, May 29, 1924.

135 MS1146. Accession 1998-M-099, Box 1, Folder 2. Mendel, L. B. Letter to H. Steenbock, June 3, 1924.

136 Steenbock, H. (1924) The induction of growth promoting and calcifying properties in a ration by exposure to light. Science 60, 224–225.

137 Schneider, H. A. (1973) Harry Steenbock (1886–1967) – a biographical sketch. Journal of Nutrition 103, 1233–1247. Steenbock lived together for many years with his parents in the family home in Madison. His father and mother died in 1942 and 1946, respectively, leaving Steenbock alone in the family home for the first time, at age sixty. The University of Wisconsin trustees finally persuaded him to accept part of the royalties from his patents; they introduced a standard policy of providing part of the royalties for other faculty members who filed patents. Upon his death, Steenbock bequeathed millions of dollars from his own estate to various philanthropies, including an academy that fostered the interests of young students in science.

138 Karrer, P., Morf, R. (1931) Zur Konstitution des β-Carotins und β-Dihydro-carotins. Helvetica Chimica Acta 14, 1033–1036; Karrer, P., Morf, R., Schöpp, K. (1931) Zur Kenntnis des Vitamins-A aus Fischtranen. Helvetica Chimica Acta 14, 1036–1040, 1431–1436; Karrer, P., Helfenstein, A., Wehrli, H., Wettstein, A. (1930) Über die Konstitution des Lycopins und Carotins. Helvetica Chimica Acta 13, 1084–1099.

139 Holmes, H. N., Corbet, R. E. (1937) The isolation of crystalline vitamin A. Journal of the American Chemical Society 59, 2042–2047.

140 Isler, O., Huber, W., Ronco, A., Kofler, M. (1947) Synthese des Vitamin A. Helvetica Chimica Acta 30, 1911–1921.

141 Dowling, J. E. (2002) George Wald (18 November 1906 – 12 April 1997) Proceedings of the American Philosophical Society 146, 432–439.

142 Wald, G. (1933) Vitamin A in the retina. Nature 132, 316–317.

143 Wald, G. (1934) Carotenoids and the visual cycle in vision. Nature 134, 65.

144 Wald, G. (1955) The photoreceptor process in vision. American Journal of Ophthalmology 40, 18–41.

145 Wald's contribution to the understanding of vitamin A and rhodopsin in the visual cycle in the retina was acknowledged in 1967 with a Nobel Prize.

Chapter 6

Milk, Butter, and Early Steps in Human Trials

The unfortunate experience of another country should make physicians in every part of the world better prepared to recognize xerophthalmia, to understand its origin and to treat or, better still, avert it. . . the history of xerophthalmia testifies anew to the enormous importance of milk as a food for children. No other article of diet can replace milk, Bloch warns us in the light of his European experience. Absence of milk from the diet, he adds, or the inclusion of improperly modified milk, is the origin of very serious diseases.

From 'Why Xerophthalmia Deserves Attention' editorial in the
Journal of the American Medical Association (1924)

Did the several findings that rats in laboratory experiments derive strength and hearty growth from some unknown factor in milk and butter have any bearing on human growth and health? Would infants and toddlers respond as lab rats did to dairy products? Even as biochemists labored to understand what F.G. Hopkins and his successors termed the 'accessory factor' in dairy products, the implications for human well-being of findings from lab experiments with animals remained uncertain [1]. Whether children whose intake of milk and butter was high would thrive and grow taller than their peers fed no dairy products had not been tested. In the case of butter, a dramatic example of its benefits was to come from Denmark through an unintentional social experiment. As for milk, the experimental model was transposed from caged rats to school-aged British orphans and impoverished boys reliant on an institution for their daily sustenance.

The High Health Cost of a Booming Dairy Industry

In the early twentieth century, Denmark had a dairy industry that was internationally regarded as exemplary. Danish butter was known for its excellent quality, and it had a strong hold on the market in Great Britain and elsewhere. Maurice Egan, a US diplomat posted to Denmark, was familiar with that country's culture and its farming, and he shared the world's high opinion of Danish agriculture. In Spring 1912, the Southern Commercial Congress in the United States invited Egan to tour Alabama, Georgia, Mississippi, Tennessee, Texas and other southern states to give talks to farmers about Danish agriculture. In his comments, he highlighted Denmark's productivity despite its lack of natural resources. Denmark, he noted, was almost entirely devoted to agriculture. 'It has no mines,' he informed his audience, 'no potential water power, no great mills. It has existed, and it seems as if it must exist, solely by means of the brain and brawn of its people applied to a soil that would be considered by the Pennsylvanians as ungrateful and in a climate which would drive a Louisianian to madness and suicide' [2]. The foundations of Denmark's robust dairy industry, according to Egan, were religion and education, a strong work ethic, the establishment of well-organized cooperative dairies since the 1880s, high standards of hygiene and sanitation, and generous government credit for farmers.

Danish butter commanded high prices even in its native country – so high, indeed, as to encourage the use of cheaper substitutes. Egan reminisced about a visit to a Danish cooking school where recipes, written on a blackboard, called for margarine in place of butter. '(T)he Danish small farmer cannot afford to use butter', Egan explained. He then went on to recount how the Danish dairy industry worked. '(M)ilk goes to the creamery to have the butter fat separated from it. The skim milk is returned to the farmer; the butter is exported for him by his cooperative society. Every pound of this butter, 'gilt-edged' in England, is too precious to be eaten by him, even at Christmas and St. Sylvester's, when he eats goose luxuriously. Hence the prevalence of margarine, well controlled by laws in Denmark, in all the cooking (recipes)'. Until the outbreak of the Great War, the British were consuming more than two-thirds of all of Denmark's agricultural exports, with Germany too importing an appreciable share [3].

As a result of butter's value as an export commodity, Denmark had the world's highest per capita consumption of margarine – forty-five pounds per person per year, compared with the Netherlands' eighteen pounds and Great Britain's thirteen (the Italians used very little margarine; likewise the French, despite the fact that margarine was a French invention) (textbox 6–1) [4]. To meet demand, Denmark had to import raw materials such as vegetable oils to manufacture margarine.

Textbox 6–1. Le beurre économique

The first margarine was made in Paris in 1869 by chemist Hippolyte Mège Mouriès. The impetus was a challenge from the French government: facing an increasing shortage of butter, officials used the 1866 Paris World Exhibition as a pretext for inviting scientists to develop an inexpensive nutritive cooking fat [5]. In response, Mège Mouriès developed a new food that he called 'margarine' for the margaric acid it contained (margaric acid is a saturated fatty acid that had been described by his compatriot, Michel Eugène Chevreul). Mège Mouriès filed patents on margarine in both France and England.

For the marketplace, the original product was called *Le beurre économique* (the inexpensive butter), but because of regulations regarding the labeling of butter, the name was changed to 'Margarine Mouriès' [6]. The Franco-Prussian war and the Siege of Paris of 1870–1871 interrupted much of the inventor's plans to profit from margarine in his own country. In 1871, Mège Mouriès sold his process to the

Dutch firm Gebroeders Jurgens, and in the following years, he sold the patents to various buyers.

Originally, margarine consisted of beef tallow and skim milk – both inexpensive ingredients – but advances in chemistry and manufacturing enabled the replacement of beef tallow with vegetable oils such as coconut oil and palm oil. At the turn of the century, margarine sold for about half the price of butter and was primarily used by poorer families. In general, the price of margarine relative to butter continued to decline in most countries during the 1900s. By the 1920s, the cost of margarine in some countries was about 30 to 40% that of butter [7] – despite the fact that, in Western Europe, the original animal fat was of domestic origin and the vegetable oils that supplanted it had to be imported.

In 1937, Denmark mandated that margarine be fortified with vitamin A. From then on, no further cases of keratomalacia were reported there. Around the same time, Great Britain and the United States also required the fortification of all margarine with vitamin A [8].

The health effect on Danes consuming little butter and much margarine became evident in 1909. At Copenhagen's Kommunehospitalet (Municipal Hospital) and the Rigshospitalet (University Hospital), infants and young children began showing up with the emaciation and dry, ulcerated eyes caused by malnutrition [9]. Some were also anemic and had diarrhea, pneumonia, and ear infections. Although xerophthalmia had been considered rare in Europe, cases of keratomalacia were now occurring at an alarming and increasing rate. Children were going blind and some were dying.

The Rigshospitalet was receiving young patients from all over Denmark; most of them with xerophthalmia had been brought to the city from rural areas. Carl Edvard Bloch (1872–1952), a pediatrician at the Rigshospitalet, had observed forty cases in Copenhagen alone [10], and additional cases were reported by a colleague at Dronning Louises Børnehospital (Queen Louise Children's Hospital) [11]. The affected children from urban Copenhagen, however, came from a single institution that housed orphans and destitute children. Searching for clues to the outbreak, Bloch undertook an investigation of the children's home.

The eighty-six children living in the institution were divided into sections A and B. Section A, with fifty-four children, housed infants and children deemed delicate or actively sick. Section B housed thirty-two healthy children who were more than a year old. Section B was subdivided into subgroups I and II, each with sixteen children; the children in Section B, groups I and II, were similar in age and health status. The hygiene conditions of the two subgroups were identical and considered adequate. None of the xerophthalmia cases occurred in section A or section B, subgroup II; all the children free of this condition were in section B, group I.

What was it, Bloch wondered, that was distinctive in the conditions of section B, subgroup I? He focused on the food.

Whole milk in one form or another provided the basis of the section A children's diet. The section B children received a breakfast of oatmeal gruel or *ollebröd* (beer-and-bread soup) and sometimes with a little whole milk; at midday, they had milk pudding made from buttermilk (i.e. the milk liquid left after churning for cheese or butter, with all the butterfat extracted), oatmeal gruel, fruit juice soup, and sometimes meat broth with barley, then fish, minced meat, and mashed potatoes, followed in mid-afternoon with cocoa, bread with margarine. And in the evening, milk pudding and bread with margarine. Butter, cream, or eggs never appeared on these children's plates, and anything cooked was prepared with margarine. The standard drink was skim – i.e. fat-free-milk. The matron in charge of section B, group I, in which all the home's cases of xerophthalmia originated, gave oatmeal gruel and dried bread to children under her care. The matron of section B, group II, gave her children beer-and-bread soup with whole milk; no cases of xerophthalmia appeared in this group.

Bloch then examined the growth curves for the children in the two subgroups. He was aware of the animal studies by Hart, McCollum, and Steenbock at the University of Wisconsin and by Osborne and Mendel at Yale University (see Chapter Five). He also was familiar with Mori's description of *hikan* – xerophthalmia – in Japanese children. The resemblance of the Danish children's growth curves and those of experimental animals on restricted diets and Japanese children in remote rural areas was striking:

> It will be seen that the children behave much as the animals in the experiments of the American physiologists. They first stop growing and then lose weight. A prolonged unvaried diet, containing either no fats or no specific lipoid bodies (as far as children are concerned, lack of fresh milk), causes both children and rats to fall sick. . . in a number of cases eye trouble develops; at first the eyes are inflamed and later become purulent and necrotic. If the diet is persisted in, both children and animals die. In both cases the dystrophy and the eye lesions can be cured by administration of cod liver oil and butter fat [12].

No further new cases of xerophthalmia appeared in the Copenhagen children's home after Bloch ordered that skim milk be replaced with whole milk throughout the institution. As for the children who were already sick, Bloch treated the infants by instituting breast-feeding when possible [13] and for the older ones, giving whole milk and cod liver oil. Even after recovery from active disease, some children with keratomalacia were left blind and with corneal scars (fig. 6.1). 'My observations. . . have made it clear', Bloch stated, 'that the disease was not due to absence of fat as such, but probably to the absence of certain bodies normally present in butter fat and in large quantities in cod liver oil. I therefore proclaimed that the disease was most probably due to "the absence of specific lipoid bodies or their constituents. . ."' Further study by Bloch showed that the young children from the countryside belonged to poor families who could not afford butter or whole milk. At best, they bought cheap skim milk, buttermilk, and margarine.

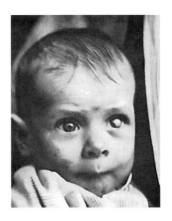

Fig. 6.1. Yrsa N., developed keratomalacia after two months' feeding with oatmeal. She was treated with cod liver oil, human milk, and cow's milk, and after recovery her right eye was nearly normal and left eye was left blind with a large central corneal scar [12]. Reproduced from the Journal of Hygiene with permission of Cambridge University Press.

Between the start of the xerophthalmia outbreak and during Bloch's examination of it, World War I had dealt a harsh blow to the Danish dairy industry – ironically, restoring the visual health of Denmark's children. The British naval blockade begun in 1915 successfully bottled up the German fleet in the North Sea, to which Germany retaliated with a U-boat offensive against merchant vessels [14]. The waters around Great Britain were now declared a war zone. Although international law prohibited indiscriminate sinking of vessels without warning and mandated the interception and inspection of merchant and other civilian ships, German commanders of U-boats and other warships did not comply with this mandate. One famous result was the sinking of the British passenger ship *Lusitania* in 1915 and the death of some two-thousand passengers and crewmen. Germany's primary objective was to cut off the Britons from imported supplies, including wheat and dairy products, and, second-arily, to deprive Denmark of raw material imports such as vegetable oils.

Thus, Denmark was stranded with no incoming vegetable oils and stuck with the dairy products it would, in peacetime, export abroad. In other words, the Danes lost out on margarine production and had to make do with its own, excellent butter – the latter, albeit, produced with increasing difficulty. Given the wartime impediments, the Danish government took control of butter in December 21, 1917, allowing adults and children an affordable weekly ration of 250 grams. The health effect was pro-found, as Bloch would note several years later. 'From that moment everyone ate butter instead of margarine and since then there has been no xerophthalmia in Denmark' (fig. 6.2a, b).

But Bloch was a realist. He predicted that, when the war ended, Denmark's manu-facturing of margarine would resume, butter subsidies and rationing would cease, and because of the cost difference, margarine would again replace butter in poor Danish households. 'It is therefore to be feared', he wrote, 'that next spring this disease of malnutrition and the disastrous ophthalmic condition will occur again amongst the children of the poorer people in Denmark' [12].

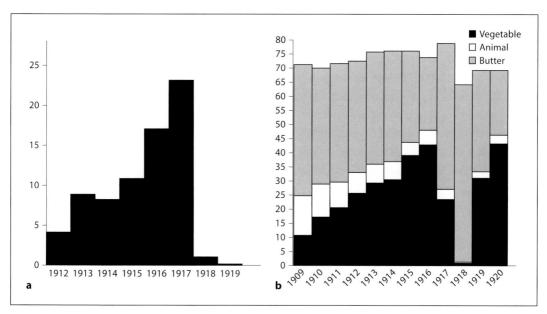

Fig. 6.2. a The rise in the number of cases of xerophthalmia admitted to the State Hospital in Copenhagen, 1912–19 [12]. **b** Butter and margarine consumption from 1909 to 1920.

With the combination of malnutrition and blindness occurring in children, xerophthalmia attracted the attention of both pediatricians and ophthalmologists. In 1919, the Ophthalmological Society of Copenhagen initiated a study under the direction of ophthalmologist Olaf Blegvad (1888–1961). Built largely on Bloch's preliminary conclusions, the study involved observation of more than six hundred cases, most of them children suffering from keratomalacia. Blegvad reviewed the historical literature for previous reports of keratomalacia in Europe and remarked, 'In older times keratomalacia in children was regarded as a most severe disease and even as a symptom of an imminent death.' The usual mortality rate before use of treatment with milk and cod liver oil, Blegvad deduced, was 50%. But physicians had learned to reduce the mortality of children with keratomalacia. 'Here in Denmark.' Blevgad wrote, 'where the milk treatment has been known and practiced for a longer time, it might be expected that the mortality from this disease should be considerably less' [15]. And indeed, Blegvad's report showed that the overall mortality rate of children with keratomalacia in Denmark was 21.5% – that is, reduced to less than half by use of treatment with milk. He further noted that Bloch's administration of milk fat and cod liver oil to hospitalized children with keratomalacia had lowered the mortality rate to about 10%.

As Bloch had foreseen, with the resumption of margarine production in 1919, keratomalacia began immediately to recur. The new cases that occurred were not so severe, however, as those encountered before and during the war. The disease became familiar to the Danish people, and their physicians had learned how to treat it [16]. In 1921, Bloch looked back on his country's experience with butter, margarine, and keratomalacia and saw the resemblance to animal experiments carried out during the previous half-century in England and the United States. Denmark, he wrote, had conducted a 'great dietetic experiment involving a whole nation' [12]. Unknown to Bloch and Blegvad, Denmark had unwittingly made a huge graphic demonstration of how vitamin A – especially the vitamin A in butter – saves the sight and lives of infants and young children.

Bloch and Blegvad's clinical observations did not go unnoticed internationally, as their reports were published in both German and English [15–17]. In 1924, the *Journal of the American Medical Association* published an editorial noting the lessons learned from Denmark and how they validate animal experimentation. The editorial was headlined, 'Why xerophthalmia deserves attention'.

Through the animal experiments of Osborne and Mendel, and of McCollum and his collaborators, in this country, attention has become directed to the phenomenon of xerophthalmia and the disease associated with it. There can be little doubt that the type of ophthalmia that can be brought about in animals through faulty diet is associated with a deficiency of vitamin A. . . clinical experience has thus fully confirmed the investigations in animals. Lest we forget in these days of antivivisection propaganda, let us bear in mind that these experiments in the laboratories pointed the way to successful therapy and prophylaxis. The unfortunate experience of another country should make physicians in every part of the world better prepared to recognize xerophthalmia, to understand its origin and to treat or, better still, avert it. . . the history of xerophthalmia testifies anew to the enormous importance of milk as a food for children. No other article of diet can replace milk, Bloch warns us in the light of his European experience. Absence of milk from the diet, he adds, or the inclusion of improperly modified milk, is the origin of very serious diseases [18].

The value of milk or butter to infants and young children was amply demonstrated in the clinical observations in Denmark. Danish physicians had demonstrated that providing vitamin A to vitamin A-deficient children reduced the severity of their illness and their risk of dying. The therapy of providing milk, butter, or cod liver oil – hence vitamin A – was established through clinical observation and from the knowledge that came from experimental animal models.

But what was the value of milk and butter to children who did not suffer from xerophthalmia?

Milk Studies in Britain: Experiments in Experimentation

Great Britain provided the setting for early questions about the general value of milk to human wellbeing, especially for school-age children. The nutrition of school-age children had become a critical issue among Britons near the end of the Boer War, when, in 1902, Major-General John Frederick Maurice sounded an alarm with his announcement that, 'out of every five men who are willing to enlist only two are fit to become effective soldiers' [19]. Concern about the apparently declining bodily condition of the British people led to formation of the Inter-Departmental Committee on Physical Deterioration. Food that was of poor quality and meager was considered the key factor in the physical 'degeneration' of the population [20]. Testifying before the committee, Alfred Eichholz, a physician and one of His Majesty's Inspectors of Schools, observed that the diet of thousands of city children was dismal. 'Their breakfasts are nominally bread and tea,' Eichholz reported, 'and the dinner nothing but what a copper (penny) can purchase at the local fried fish shops, where the most inferior kinds of fish are fried in reeking cotton-seed oil' [21]; milk, fruits, and fresh vegetables were virtually absent.'

Supporting Eichholz's point, a large proportion of schoolchildren were found to be malnourished, and a new commission, the Committee on the Medical Inspection and Feeding of Children, was appointed in 1905. It recommended that the government provide meals through the schools 'to feed the most destitute children regularly, rather than a large number irregularly' [22]. In 1906, the Provision of Meals Act was passed, enabling local school authorities to subsidize free or low-cost meals to school children [23]. The financial position of the local school authorities was strengthened (although not until 1914) with a new Provision of Meals Act; this legislation raised cost limits and allocated monies to support the provision of meals in schools [24].

Although the biochemical experiments of the past half-century demonstrated that milk was important to lab animals' growth, and experience strongly suggested that young humans benefitted from it, providing children with adequate cows' milk presented problems. Poor hygiene and bacterial contamination of milk set off epidemics of diarrhea and typhoid fever [25]. But the advent between 1910 and 1920 of wider controls and inspection of the dairy industry brought about safer milk as a result of improved standards of hygiene, pasteurization, and proper refrigeration during transport and storage. The public perception of milk, which had seen by some Britons as potentially dangerous, was largely rehabilitated during the 1920s and 1930s through a combination of applied science and enforcement of government regulations concerning animal housing and feed, veterinary inspections, testing cows for tuberculosis, and monitoring of bacterial counts in milk. Together, growing public confidence in the safety of cows' milk [26] and the availability of dependable and safe supplies allowed for scientific evaluation of milk in children's diets.

In 1921, amid growing concern about the adequacy of the diet of English school children, especially children from relatively poor industrial areas, the Food Department of the Ministry of Health took steps to study the effect of fresh, pasteurized cows' milk on young school-age boys. Although the significant benefits of milk given to ill-nourished children were recognized, whether milk added to the basic diet of relatively healthy children also fundamentally improved their nutrition was still unproven. In May 1921, the Accessory Food Factors Committee of the Ministry of Health reviewed and approved a study proposal titled, 'Scheme for the Study of the Effects Produced in the Growth and Nutrition of Underdeveloped Children by Addition of Milk to the

Fig. 6.3. Boys assembled in front of Saturday House, one of the typical cottages at Boys' Garden Village, Woodford Bridge. Reproduced with permission of www.history-in-pictures.co.uk.

Standard Diet' [27]. The committee included three scientists noted for their familiarity with the benefits of milk: physiologist-physician Charles James Martin (1866–1955), who was director of the Lister Institute of Preventive Medicine, F.G. Hopkins, and Professor Jack Drummond. Hopkins, for whom such a study would be a natural sequitur to his own work on 'accessory factors', provided the main scientific guidance for the design of the investigation. To carry out the experiment, the council engaged Harold Corry Mann, a pediatrician at London's Evelina Hospital for Sick Children.

Corry Mann proposed that the study be conducted at the Boys' Garden Village in Woodford Bridge, one of the many orphanages run by a large institution called Dr. Barnardo's Homes that cared for destitute and homeless children. The Boys' Garden Village, which usually had about five to six hundred boys in residence at any one time, occupied a large estate northeast of central London. It had nineteen cottages, each of which housed between thirty and forty boys (fig. 6.3). The dormitories in the cottages were spacious and well provided with modern sanitary facilities. The Boys' Garden Village had its own school, a large dining hall with adjacent kitchens, a bakery, a small hospital staffed by nurses, and a swimming pool. The institution fed, clothed, and educated the boys until they were required either to emigrate to work in the colonies, to be transferred to serve in the navy or the merchant marine, or to take up apprenticeships in trades. The committee overseeing the study concurred with Corry Mann that the Boys' Garden Village was ideal.

The study commenced in November 1921 when three cottages at Boys' Garden Village became available for occupancy. (Unbeknownst to Great Britain's Ministry of Health or Corry Mann, a somewhat similar study was already in progress in the

United States when Corry Mann began his own; the results of the US study were never to appear in the peer-reviewed scientific press; textbox 6–2.)

According to plan, the experimental subjects, all between ages seven and eleven and weighing forty-five to sixty-five pounds, were to be 'typical of the average London boy found on the south side of the River Thames' with 'mouse-colored hair and eyes with grey or grey-blue iris' – hence, 'no boys of colour or of foreign, Latin, Scandinavian, or Hebrew type. . .' [28]. To qualify for the study, the boys had to be in good health, without signs of rickets, infections, or other problems. (By present-day standards, about 25% of the boys selected for Corry Mann's study arrived at Boys' Garden Village underweight – i.e. malnourished. More than 40% were stunted, which reflects chronic malnutrition. The proportion of malnourished or stunted children cited here is similar to that seen today in some developing countries [29].) Corry Mann divided them into three groups that were alike in weight and age, and assigned the groups to the three cottages – named, somewhat whimsically, 'Christine', 'Natal', and 'Lucking'. Experimental diets were to be assigned by cottage.

Textbox 6–2. An echo in Baltimore

The trials conducted at the Boys' Garden Village were not the first controlled studies of supplementary feeding with milk. In 1919, Elmer McCollum at the Johns Hopkins School of Hygiene and Public Health conducted a study with boys and girls from a Baltimore orphanage. His explicit goal was to determine whether the conclusion he and many colleagues had reached regarding milk in experimental animals could be applied to the growth of children [30]. McCollum regarded the source of his subjects as ideal, being, as he described it, 'an institution from which there were two-hundred and thirty-six negro children ranging in age from early infancy to twelve years, whose environment and food supply left little to be desired in relation to our objective'. Most of the children in the Baltimore institution were malnourished. McCollum selected for the study eighty-four children ranging in age between four and ten. After matching by age, size, and condition, he divided them into two treatment groups of forty-two: one group received one quart of milk per day; the other, no milk.

McCollum was impressed by the effect of milk on the children's growth. 'The results of the supplementing of the institutional diet with milk were so striking that after 15 months it was decided to supplement the diet of the check [control] group with a quart of reconstituted milk per day, in order to demonstrate whether these children, at least a considerable number, possessed the capacity to grow which they were not manifesting on the institutional diet. As was anticipated, during the next six months a considerable number of the control group grew at rates comparable to what had been observed in the original milk group.'

As for the milk-fed boys' behavior, Corry Mann and McCollum unwittingly made similar and nearly contemporaneous observations. Corry Mann's description succinct: '(T)hey became far more high-spirited and irrepressible, being often in trouble on that account.' McCollum was more discursive: 'There was likewise a marked change in the behavior of the milk-fed group as contrasted with the check group. That latter were apathetic and very tractable. The discipline of the institution was strict and these children were all quite obedient. Those in the milk-fed group, on the other hand, soon caused annoyance to their teachers by their restlessness and desire for activity. They also were frequently guilty of infractions of the rules.'

McCollum's milk study did not become widely known. Instead of reporting his results promptly in a scientific journal, he published them in the 1924 proceedings of the World's Dairy Congress held the year before in Washington, D.C. Whatever his reasoning, McCollum also refrained from mentioning his milk study in his autobiography.

Corry Mann screened and selected young boys who had recently arrived, as he put it, 'from the slums' of London [28].

For a preliminary four-month period, Corry Mann established a baseline by observing the three groups closely in order to characterize the institution's basic diet and to determine whether the organization and administrative routine of the dining hall would allow different dietary interventions for boys in separate cottages. He concluded that the institution's organization and routine would be conducive to the proposed study. During this period, he also noted that the boys spent a great deal of time out of doors exposed to the sun [31].

The basic Boys' Garden Village diet was typical of what large institutions caring for children fed their charges, and Corry Mann meticulously documented it. Meals consisted largely of beef, pork or ham, or fish; potatoes; beans; bread; margarine; cocoa; jam; and rice pudding. Steamed cabbage and reconstituted dried peas were served once or twice a week. In addition, the boys received cheese once a week and fresh fruit about once a month. Except for tinned spinach perhaps once a year, dark leafy green vegetables were almost never served; carrots appeared about once a year [28]. Corry Mann noted with some satisfaction that, with regard to the study, 'the diet is a splendid one. . . low in caloric value and also in items containing 'fat-soluble [A]' [32]. He observed, in addition, that many of the boys are 'a stone underweight and six inches under height. . .. The more that I see of the "basal diet" of all the Homes, the less I think of it, which makes it all the better from the standpoint of accessory food factors' [33] (textbox 6.3 for a modern nutritional analysis).

Textbox 6–3. A modern analysis of the vitamin A in Corry Mann's study diets

Corry Mann's detailed descriptions of the meals served at the Boys' Garden Village can be analyzed by modern methods to assess the amount of vitamin A the boys were receiving in the basic diet and with various food items added. The basic diet contained about 63 μg retinol activity equivalent (RAE) of vitamin A per day, which can be compared today with the Institute of Medicine, Food and Nutrition Board recommendations of an Estimated Average Requirement (EAR) of 275 μg RAE per day for a boy aged 4–8 years and 445 μg RAE per day for a boy aged 9–13 years. The EAR is defined as the nutrient intake estimated to meet the nutrient requirement of half of healthy individuals of a particular age and gender group. The Recommended Dietary Allowance (RDA) for boys of these two age groups is 400 μg RAE and 600 μg RAE per day [34]. In regard to vitamin A, the diet in the Home contained about 16% of the RDA for the younger boys and 10% of the RDA for the older boys. Obviously, the basic diet at the Home was grossly deficient in vitamin A, as consistent with Corry Mann's initial impressions that the diet was low in 'fat-soluble A.' (see table 6.1 for the daily intake of vitamin A in each of the seven different treatment groups.) Only the boys in the milk and the butter group approached or exceeded the average requirement for vitamin A. Although the ration of watercress was intended to represent 'fat-soluble A' alone, the amount of watercress was not sufficient to provide much of any increase in vitamin A in the diet of the boys. In addition to the 67 μg RAE already in the basic diet, it would have required 200 g of watercress to provide 333 μg RAE for a boy aged 4–8 years and 320 g of watercress to provide 533 μg RAE for a boy ages nine to thirteen years to reach 100% of the RDA.

Providing the required vitamin A through watercress alone would have been a difficult proposition, as the younger and older boys would have needed to consume six or nine cups of watercress per day, respectively [35]. The difficulty in meeting the daily requirements for vitamin A through dark green leafy vegetables alone is well illustrated in this study, as most school-aged children probably would refuse to eat such a large amount of watercress every day for a year.

In spring 1922, when Corry Mann's observation period ended, the boys in the three cottages were assigned to receive specific diets: either (1) the basic diet, (2) the basic diet plus one pint of milk per day, or (3) the basic diet plus three ounces of sugar per day. Although the institution had a communal dining hall for the boys from all nineteen cottages, the boys in Corry Mann's experimental groups were seated at separate tables to ensure that each group received its correct food allowance and nothing different. Average food consumption was recorded by weighing the food served at each table before eating began and the leftovers after the meal was finished. The boys were weighed every two weeks and their height was measured every three months. After a year, the investigators and workers at the home could not help noticing that the boys receiving the daily one pint of milk had grown taller and gained more weight than the boys on either of the other two diets. The boys who received milk every day had an obvious advantage in growth.

Table 6.1. Vitamin A content of the diet in the seven different treatment groups at the Boys' Garden Village

	Control	Milk	Sugar	Butter	Margarine	Casein	Watercress
Basic diet, µg RAE/day	63	63	63	63	63	63	63
Addition, µg RAE/day	none	173	0	339	0	0	47
Total, µg RAE/day	63	236	63	402	63	63	110
RDA for boys 4–8 years, %	16	59	16	100	16	16	27
RDA for boys 9 years and older, %	10	39	10	67	10	10	18

The Recommended Dietary Allowance (RDA) for boys ages 4–8 years is 400 µg RAE/day and for boys ages 9–13 years and older 600 µg RAE/day [34].

Interference from Within and Without

Corry Mann requested additional time to accumulate more data. But the secretary of the Medical Research Council, Sir Walter Morley Fletcher (1873–1933), was concerned about the study's cost and requested that Corry Mann seek a definitive decision from a trained statistician as to the results so far. Sir Walter had an additional worry [36]: 'I understand from you at the same time that, owing to the striking nature of some of your results, the nurses at the Home were beginning to spoil your controls from excessive sympathy with the boys kept on the original diet without supplement. This in itself seemed likely to call for reconstruction of the programme.' So the direction of the experiment was changed in mid-course, and Corry Mann consulted with Hopkins to see what additional questions could be answered.

The new plan called for another year of continuing two groups' regimens as before: the basic diet alone and the basic diet plus milk; for the third group, the sugar addition to the basic diet was dropped and replaced with one and three-quarters ounces of butter per day [37]. The experiments of McCollum, Davis, Osborne, Mendel, and others had all shown the potency of butter in the growth of small rodents, but butter's influence on human growth had not be scientifically studied. In autumn 1924, Arthur W. J. MacFadden, director of the Food Department of the Ministry of Health, visited Boys' Garden Village and urged the Medical Research Council to continue funding for at least another year. 'I cannot imagine a more favorable field, or more favorable circumstances for experiment and observation of this kind than happily exists there. . .' [38].

During winter 1923–1924, Corry Mann noted 'a complete absence of illness among the boys with the milk ration', whereas the illness rates were relatively higher in the other Boys' Garden Village cottages because of an outbreak of influenza and a flare-up of measles and scarlet fever. He also noted [28], 'After eighteen months' observation a casual visitor entering the dining hall. . . would never fail to recognize the table of that house which was alone receiving the extra ration of milk, the boys of that house being obviously more fit than those of any other house. In addition, they became far more high-spirited and irrepressible, being often in trouble on that account. . .'.

The following year, Corry Mann requested to add another treatment group to the investigation: he and Hopkins wanted to 'test the milk ration against a ration of watercress, (for its) fat soluble A, (with) the boys on basic diet remaining as the control.' He felt pressed to begin this next phase promptly, since he was afraid that the authorities overseeing Dr. Barnardo's Homes would alter the resident boys' basic diet. 'They are already aware that the addition of one pint of uncooked milk to the basic diet has immensely improved the physical condition of the boys who are receiving the ration', Corry Mann wrote. 'The work is obviously incomplete until the milk has been tested against an item of diet which will provide fat soluble A in the raw state without the additional protein, fat, and carbohydrate present in the milk ration. . .' [39].

The Medical Research Council and Corry Mann discussed publishing the results that showed the effects of milk on growth of schoolboys, but they were worried. If they did so, it would lead to 'immediate cessation of the trials, because the Barnardo people could not continue to keep the "control" boys any longer without the additional nourishment that this work has proved to be so valuable' [40]. So, publication of the results was deferred and the milk trial continued. A new group of boys in cottages called 'Britannia' and 'Empire' were assigned to receive the basic diet plus watercress in place of the supplementary milk, in order to determine whether the fat-soluble A in watercress would yield the same growth benefit as had been seen with milk. Hopkins expressed strong support to continue the investigations:

> The thousands of lads that have been sent to the Colonies and elsewhere from these institutions have therefore been definitely under-nourished and stunted. I, for one, have suspected this, but I did not *know* it. The results show that even when the youngsters are fairly well satisfied, and show no tendency to join in Oliver Twist's revolutionary demands, they may yet be undernourished. . .. I took little interest in Mann's work until I suddenly realized how exceptional are the conditions he has obtained for making *real* comparisons. Each house a nutritional unit – full freedom to control and watch the feeding – each unit a group of 30 boys! [41].

In early summer 1924, Secretary Fletcher visited the ongoing study at the Boys' Garden Village and came away with mixed feelings. He professed 'complete admiration for the excellence of the opportunity which has been offered there for studies of this kind', but at the same time, 'great anxiety lest this very great opportunity should have been inadequately used.' Although Corry Mann had stayed in close consultation with Hopkins regarding the study, Fletcher was worried: might the trials have been planned better? [42]. Fletcher's concern was that what began as a study of the efficacy of milk in promoting growth and good health had metamorphosed into a study dealing with the 'vitamin question'. He sought the views of two eminent colleagues, Professor D. Noël Paton (1859–1928), a physiologist at Glasgow University, and Jack Drummond [43].

Paton's review of Corry Mann's work, stated in characteristically unsparing terms, went far in corroborating Fletcher's apprehension. Paton's position, conveyed in a letter to Fletcher, was that 'no new results' had been shown despite two years of work, and that the results 'simply demonstrate the well-known fact that a substantial addition of proteins, fats, and calcium to an admittedly unsatisfactory diet will improve nutrition' [44]. Fletcher took a paternal tone in his reply [45]: 'I think you are inclined not to realize that human trials of this sort... cannot be conducted like animal experiments, simply because of the time limit and the impossibility of multiplying simultaneous experiments. Beyond a certain point with human cases, especially if successful results are got, the control is bound to break down, because the humanitarians begin to say that the controls are being "starved", although before the trials everyone was perfectly satisfied with the basic diet'. Months later, Fletcher again reflected on the progress of the milk study and declared to the Medical Research Council [46], 'The first stage has been won. An institution diet commonly used has been proved lacking and the addition of a pint of milk beneficial. The next stage, the analysis of the beneficial effect, is now going on. Mere increase of calories by addition of sugar has done nothing, and it is worthwhile providing this to the lay mind by such a "field" experiment. Addition of butter has been tested and found less good than milk. Watercress is now being tested from the point of view of vitamines'.

Corry Mann's nutritional trials did not end at watercress or because of any tension about applied and pure research. The evacuation of a fifth cottage – 'New Zealand' – at the Boys' Garden Village enabled Corry Mann to test the results of adding casein to the basic diet.

At the urging of Jack Drummond, yet a seventh treatment group was added to the study, and another dietary component was tested: margarine. A local margarine factory agreed to provide their vegetable margarine for testing in the trial. Corry Mann noted with some satisfaction that the margarine had a composition of 'vitamin A *absent,* containing *no* butter-fat at all' [47]. As before, Hopkins approved, now taking a patriotic view: Corry Mann was responding to 'national question' as to whether substituting margarine for butter might in fact be deleterious to the wellbeing of Great Britain's growing children.

The nutrition experiments at the Boys' Garden Village finally came to a close at the end of 1925 (Corry Mann wanted to close down the study as quickly as possible after he requested, but was refused, a permanent position with the Medical Research Council). A few months later, the council distributed a final internal report to various scientists. It circulated with independent statistical verification: Major Greenwood, a respected statistician who served the Ministry of Health, provided a statistical analysis. Greenwood presented the results by first grouping the boys according to their original weights – 'forty-pounders', 'fifty-pounders', and 'sixty-pounders'. His results generally confirmed Corry Mann's impression: the boys in the milk and butter groups showed significantly greater gains in weight and height than did those in the basic diet group [48] (see textbox 6–4 for a modern re-interpretation of the study results.)

Textbox 6–4. A retrospective look and re-analysis of the milk study

The work carried out at the behest of the Medical Research Council can be seen as exploration in the virtually uncharted territory of human trials. The deliberations among the scientists involved signal the newness of the kind of work Corry Mann was conducting. Even more striking is the fact that the investigators kept adding intervention groups to the study as it went along – a practice that would be most unusual today and probably frowned upon.

The design and execution of the milk study would be very different if undertaken today, and early trials cannot be judged by the modern standards of properly controlled clinical trials. Certain entities, principles, and practices – review of protocols by institutional review boards and ethicists, written informed-consent procedures, random assignment of participants to treatment groups, use of placebos, and monitoring of studies by outside committees – have become standard practice in the world of experimentation involving humans.

Likewise, the original raw data published in Corry Mann's report can be analyzed quite differently and in greater detail today. All boys were in the study for at least twelve months, while some were in a treatment group for a longer period. To keep the comparisons between the seven groups consistent, present-day analyses are limited to the first twelve months. Using modern criteria and definitions for underweight and stunting, no significant differences appeared between the boys in the seven different groups at baseline (table 6.2). Today, stunting, reflecting long-term chronic malnutrition in children, is often used as the outcome in clinical trials of nutrition interventions. After twelve months of treatment, the changes in height and weight and other growth indices are shown in table 6.3. At the twelve-month follow-up visit, there were significant differences in both changes in weight and in height from baseline among the seven groups, with boys in the milk and butter groups showing the largest gains in weight and height. The proportion of boys who were stunted at baseline and twelve months later is shown in figure 6.4. Again, the largest reductions in stunting from the time of enrollment in the study to the twelfth-month follow-up visit were seen in the boys who received either milk or butter, while stunting was virtually unchanged in boys who received watercress, casein, margarine, sugar, or no intervention (control).

Solid though they were, Corry Mann's results met with some professional grumbling and unfortunate commercial enthusiasm. The nutritionist Harriette Chick, who conducted the pioneering studies of rickets in Vienna (chapter 5), was asked to evaluate the study. She lamented [48]: '(My) criticism comes too late. I think a group should have had cod-liver oil, i.e. addition of vitamins A and D without extra calories. Of course the watercress addition gives vitamin A without extra calories and a year

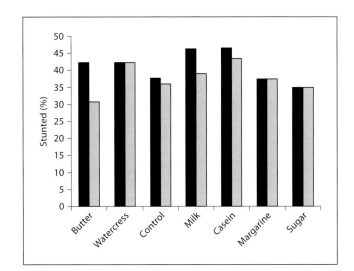

Fig. 6.4. Change in stunting in the different intervention groups: the greater reduction in stunting occurred in the butter and milk groups. Baseline, black bars; twelve-months follow-up, grey bars.

Table 6.2. Anthropometric measures of schoolboys at baseline in the milk trial at Boys' Garden Village, Woodford Bridge, London

Characteristic[1]	Butter	Watercress	Control	Milk	Casein	Margarine	Sugar	p
Number	26	26	61	41	30	16	20	–
Age, months	121 (18)	110 (17)	111 (18)	116 (18)	114 (16)	113 (15)	118 (19)	0.21
Weight, kg	26.4 (4.3)	24.2 (3.7)	24.3 (3.7)	25.0 (4.1)	25.4 (4.3)	25.0 (3.4)	25.7 (3.5)	0.33
Weight-for-age Z-score	−1.40 (0.72)	−1.37 (0.82)	−1.41 (0.77)	−1.53 (0.95)	−1.32 (1.05)	−1.27 (1.15)	−1.37 (0.78)	0.93
Underweight, %[2]	26.9	15.4	22.9	29.2	30.0	12.5	30.0	0.69
Height, cm	126.8 (4.3)	122.9 (8.2)	123.2 (8.9)	123.8 (8.8)	124.2 (9.8)	124.2 (7.5)	125.6 (8.8)	0.54
Height-for-age Z-score	−1.88 (0.80)	−1.82 (0.92)	−1.83 (0.87)	−2.07 (0.85)	−1.86 (1.07)	−1.75 (1.29)	−1.90 (0.77)	0.77
Stunted, %[2]	42.3	42.3	37.7	46.3	46.6	37.5	35.0	0.95

[1]Mean and SD shown for continuous variables.
[2]Underweight and stunted are defined as weight-for-age <−2 SD and height-for-age <−2 SD, respectively.
The US National Center for Health Statistics child growth standards were used to calculate weight-for-age (a measure of underweight) and height-for-age (a measure of stunting). A Z-score is equivalent to one standard deviation. Thus, for a boy who has a Z-score of −1.80 for weight-for-age is 1.8 SDs below the average growth for boys of the same age. For further details on child growth standards, see de Onis [29].

Table 6.3. Weight and height in the different treatment groups at the 12 month follow-up

Characteristic[1]	Butter	Watercress	Control	Milk	Casein	Margarine	Sugar	p
Number	26	26	61	41	30	16	20	–
Weight, kg	29.3 (4.6)	26.7 (3.9)	26.0 (3.7)	28.2 (4.3)	27.1 (4.4)	27.4 (3.3)	27.9 (3.7)	0.04
Change in weight, kg	2.8 (0.6)	2.4 (0.4)	1.7 (0.5)	3.2 (0.5)	1.8 (0.5)	2.4 (0.4)	2.2 (0.7)	<0.0001
Weight-for-age Z-score	−1.39 (0.63)	−1.38 (0.77)	−1.61 (0.74)	−1.37 (0.83)	−1.51 (0.96)	−1.30 (0.99)	−1.49 (0.76)	0.58
Underweight, %	23.0	15.4	24.6	26.8	26.6	12.5	35.0	0.69
Height, cm	132.5 (8.4)	127.3 (8.2)	127.9 (8.5)	130.5 (8.2)	128.7 (9.3)	128.9 (7.0)	130.6 (8.3)	0.23
Change in height, cm	5.7 (1.2)	4.3 (0.7)	4.6 (0.9)	6.7 (1.1)	4.5 (0.8)	4.6 (0.8)	4.9 (1.3)	<0.0001
Height-for-age Z-score	−1.74 (0.76)	−1.88 (0.89)	−1.84 (0.81)	−1.75 (0.72)	−1.88 (0.99)	−1.73 (1.16)	−1.87 (0.67)	0.96
Stunted, %	30.7	42.3	36.0	39.0	43.3	37.5	35.0	0.96

[1]Mean and SD shown for continuous variables. For changes in weight and height in the different groups, linear regression models were adjusted by age and baseline weight and height, respectively. χ^2 tests were used to compare categorical variables across the seven groups.

ago we believe it would also give vitamin D. So hard to plan a large experiment when the subject is moving so fast.'

The United Dairies Limited requested an advance copy of the report, noting that they had supplied a great deal of the milk used in the study. Corry Mann informed them with regret that he did not have the authority to send any advance copies of his report [49].

After a delay caused by a printers' strike, however, the Medical Research Council finally published the results of his study on August 4, 1926. The findings, presented under the title 'Diets for Boys during the School Age', showed that generally healthy boys who received milk or butter as part of their regular diet grew better in both weight and height than did boys receiving no milk (fig. 6.5, 6.6) [28]. The addition of sugar, protein (casein), or margarine to a basic diet did little to improve the boys' growth in comparison to milk. The boys who received watercress gained more weight than did the boys who received the basic diet only, but there were no differences in average height increase between the with- and without-watercress groups. Corry Mann's results were quickly disseminated by scientific journals, with summaries reported in the *Lancet* and the *British Medical Journal*.

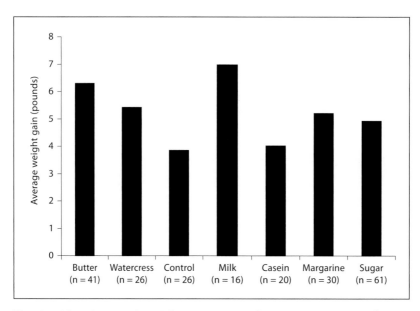

Fig. 6.5. Mean increase in weight over one year by treatment group in the Boys' Garden Village, Woodford Bridge, London.

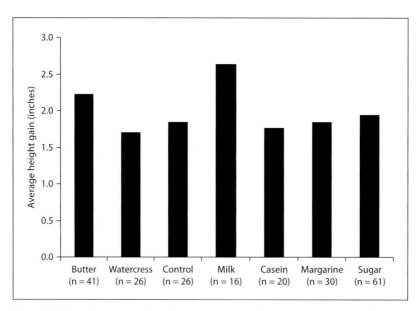

Fig. 6.6. Mean increase in height over one year by treatment group in the Boys' Garden Village, Woodford Bridge, London.

To illustrate the Medical Research Council's Report on the Diets for Boys during School Age.

These figures represent groups of boys who were given an ordinary diet for a year. At the end of that period six groups were given the extras as shown. The average annual gain in weight and height of boys given a pint of milk daily was 6.98 lbs. and 2.63 ins. respectively, whilst the boys given no extras gained only 3.85 lbs. and increased in height only 1.84 ins.

Fig. 6.7. Milk advertisement based upon Medical Research Council study. Courtesy of the National Archives, Kew, London.

Industry was quick to react. In Washington, D.C., the secretary of the Institute of Margarine Manufacturers wrote to the Medical Research Council in London to complain [50]: 'You can readily see that the dairy and butter interests in this country are more interested in destroying our market for margarine than they are in promoting public health'. If the American margarine lobbyists saw the cup as half empty, Great Britain's dairy promoters saw it as half full. The National Milk Publicity Board was excited about the study results and requested many copies to distribute. But in November 1927, C. Killick Millard, Medical Officer of Health for the City of Leicester, wrote to the Medical Research Council with a complaint of his own: the National Milk Publicity Council was distributing misleading advertising (fig. 6.7). 'It purports', Millard wrote [51], 'to illustrate the Medical Research Council's Report... it seems to me that the postcard is inaccurately drawn and is in consequence very misleading... it is obvious, even from a casual glance, that the boy labelled "Sugar" is taller than the boy labelled "Butter" while the boy labelled "Watercress" is the smallest of all, and markedly smaller than the boy labelled "No Extras". This, of course, is quite wrong according to the figures given in the report.'

For its part, the Medical Research Council reprimanded the Milk Publicity Council for the misleading advertisement [52]: 'From the text printed under the illustration, however, it might justifiably be inferred that the illustration shows the average results of feeding boys, originally all of the same size, for one year on the respective diets. On this basis the proportions shown are of course absurd. The average difference between a boy with added milk and a boy on the basic diet would be only about three-quarters of an inch (big enough indeed for one year), and not about a head as figures.'

The council conveyed the opinion that such exaggerated claims would cast suspicion on the science itself. In response, the milk council apologized for the advertisement with a somewhat unpersuasive explanation [53]: 'The fact that the boy marked Sugar is apparently taller than the boy marked Butter is due to a photographic error. . .. (T)he Sugar boy is slightly out of alignment which places him nearer the camera.'

In commissioning the milk study, the Medical Research Council could not have intended to exacerbate the inherent tension between research and the desire of industry to enrich itself by disseminating findings through advertising. But this is surely what happened with Corry Mann's results – made worse by the fact that the information circulated as the basis of industry's claims was inaccurate.

Great Britain's milk experiments continued. A large trial soon followed that involved 1,425 school children from seven centers in Scotland and Ireland (Aberdeen, Belfast, Dundee, Edinburgh, Glasgow, Greenock, and Peterhead). Under supervision in the schools, children received – as supplementary feeding – either: (1) whole milk, (2) skim milk, (3) a biscuit (meant to provide the energy equivalent of skim milk), or (4) nothing, as a control group. After seven months of treatment, the children receiving milk showed greater increase in height and weight than the children receiving a biscuit or nothing. There were no significant differences between the two milk groups [54]. The period of investigation was extended, with somewhat similar results, except that the longer follow-up suggested that children in certain age groups had better growth on whole milk than skim milk [55].

A larger, quite ambitious trial involving twenty thousand school children was conducted in Lanarkshire, Scotland. Children in this study received as supplementary feeding in the schools either raw whole milk, pasteurized whole milk, or no milk [56]. Children who received either raw or pasteurized milk overall showed greater increases in weight and height than did the children who received no milk. But William S. Gosset (1876–1937), a statistician who used the pen-name 'Student', criticized the design of this study and suggested improvements for future experiments [57]. Although 'Student's' analysis corroborated that the study showed that milk increased the height and weight of the children, Gossett pointed out that the measurements were imprecise since the children had been weighed wearing their indoor clothing. Furthermore, children had been assigned to treatment groups without any matching by weight, height, sex, and age; thus, there was no assurance at the start of the study that the groups were comparable. Nor was assignment of children to treatment groups done randomly, and he suggested that it should have been done by a toss of a coin [58].

Lessons Learned

Critics in some quarters have asserted that the Great Britain's school milk programs were meant to benefit the dairy industry more than the schoolchildren [59], disregarding the reality that the diet of British schoolchildren during the period of the studies was otherwise extremely poor in vitamin A. When Corry Mann and other scientists conducted their milk trials, the prevalence of chronic malnutrition was high not only in Great Britain but also in the United States. Studies from the US National Health and Nutrition Examination Survey (1999–2002) confirmed that milk consumption is significantly associated with children's growth [60]. And studies from developing countries, where malnutrition rates today are comparable to that of Great Britain in the 1920s and 1930s, show that consumption of cow's milk is significantly associated with the height that growing children reach [61].

Magazine articles and popular books on vitamins have emphasized the value of butter as a source of vitamin A. In *Vitamins and Your Health* (1936), Marguerite Gauger retold the lesson from Denmark, with a comment on the role of government:

How close the relationship between vitamin A and blindness can be was illustrated in a remarkable manner by the Danish Government a few years ago. During the World War the ravages of xerophthalmia appeared in all parts of Denmark. An investigation revealed that the children of the poor were the chief sufferers. . .. The investigation thus disclosed that although Denmark was world-famous for its butter and milk, its own children were being deprived of these essential dairy products. . .. Nature cannot thus be fooled. Deprived of the proper vitamins to build healthy young bodies, she visited the innocent children with sightlessness. As soon as the war was over, the Danish Government stepped into the situation. It may be rugged individualism to eat what one chooses, and it may be paternalism or Socialism for a government to dictate a diet to its citizens. The Danish Government did not care. It was concerned with the health of little Danish children more than it was interested in political shibboleths. 'Butter must form an integral part of the diet of every child' it commanded, with a significant warning of severe punishment for those who dared to disobey. Xerophthalmia has been practically wiped out in Denmark [62].

In Great Britain, the results of the milk experiments lent strength to efforts to provide school children with milk. Under the 1906 Education (Provision of Meals) Act, the government sought to assure that lack of food prevented no school child from taking full advantage of the education provided [63]. Many schools therefore provided both milk and cod liver oil to their disadvantaged students – the latter to ensure that poor children were receiving sufficient vitamin A. In 1932, supplementary milk reached 900,000 children [64]. By 1938, 'the last full year of peace' between the end of World War I and the outbreak of World War II, 2.5 million British schoolchildren were receiving milk, 560,000 of them at no cost to their families [65].

In the United States, the American Medical Association's Council on Foods and Nutrition recommended in 1939 that margarine be fortified with vitamin A [66]. Federal subsidy of milk for school children began in 1940 in Chicago and New York. In the early 1950s, skim milk, which lacked vitamin A, was fortified with 2,000 IU of vitamin A per quart [67]. The Special Milk Program was authorized in 1954 to encourage consumption of fluid by serving milk at the lowest possible cost or at no cost at all for eligible students. In 1961, the Food and Nutrition Board of the National Nutrition Council and the Council on Foods and Nutrition of the American Medical Association reaffirmed their endorsement that margarine, fluid skim milk, and dry nonfat milk should all be fortified with vitamin A [68]. The Special Milk Program became part of the Child Nutrition Act of 1966. Milk consumption in the schools increased more than tenfold between 1946–1947 and 1969–1970, from 228 million half pints of milk served to 2.7 billion half pints served under the Special Milk Program of the Child Nutrition Act [69].

Even as the fundamental necessity of milk to the growth of children was proven and laws were passed to mandate the inclusion of milk in children's diet, problems recognized in the late-nineteenth century as arising from vitamin A deficiency still plagued children and haunted scientists. In Europe, the severe vitamin A deficiency Edvard Bloch observed in Denmark was matched by reports from elsewhere in Europe and across the Atlantic. German ophthalmologists Albrecht von Graefe [70] and Julius Hirschberg [71] saw children with corneal ulceration, keratomalacia, and blindness. In Great Britain, Sydney Stephenson [72] and James Calvert Spence [73] too noted children with eye disease. And from the United States came the similar reports of Kenneth Blackfan and S. Burt Wolbach [74]. These clinical cases were dramatic and raised the question: are they just the tip of an iceberg of pathologies resulting from vitamin A deficiency?

And if many children presented physicians with Bitot's spots and other symptoms known to be associated with vitamin A deficiency, might there not be hundreds more who were also affected by vitamin A deficiency but showed no overt clinical signs? The notion of subclinical vitamin A deficiency became a subject of discussion in major medical journals in the 1920s. On the basis of observations of humans in Denmark and the animal studies of other scientists, Erik Widmark (1889–1945), a professor of medical and physiological chemistry at the University of Lund in Sweden, concluded, 'There must be in a population in which xerophthalmia occurs a much larger number of cases in which the deficiency in vitamin A, without producing the eye disease, is the cause of a diminished resistance to infections, of general debility, and of malnutrition' [75] (chapter 7).

A state of subclinical vitamin A deficiency was acknowledged as 'the borderline between health and disease,' in which a child might appear healthy, but, in the face of infection, could fare badly because of underlying vitamin deficiency [76]. Stuart J. Cowell, Professor of Dietetics at the University of London raised an essential question: 'Can a liberal supply of vitamin A affect the susceptibility to infection of individuals who show no signs of any deficiency of vitamin A?' [77].

References

1 Hopkins (1912); McCollum & Davis (1913); Osborne & Mendel (1913).

2 Egan, M. F. (1913) Notes on agricultural conditions in Denmark which served as a basis for the Hon. Maurice Francis Egan's series of lectures delivered in various southern states in the spring of 1912 under the auspices of the Southern Commercial Congress. US Senate, 62nd Congress, 3rd Session, Document No. 992. Washington, Government Printing Office.

3 Salmon, P. (1997) Scandinavia and the great powers 1890–1940. Cambridge, Cambridge University Press, p. 23.

4 Snodgrass, K. (1930) Margarine as a butter substitute. Fats and Oil Studies of the Food Research Institute, No. 4. Stanford, Stanford University Press, p. 180.

5 Hoffmann, W. G. (1969) 100 years of the margarine industry. In Stuyvenberg, J. H. van (ed.) Margarine: an economic, social and scientific history 1869–1969. Liverpool, Liverpool University Press, pp. 9–33.

6 Alphen, J. van (1969) Hippolyte Mège Mouriès. In Stuyvenberg, J. H. van (ed.) Margarine: an economic, social and scientific history 1869–1969. Liverpool, Liverpool University Press, pp. 5–7.

7 Tousley, R. D. (1969) Marketing. In Stuyvenberg, J. H. van (ed.) Margarine: an economic, social and scientific history 1869–1969. Liverpool, Liverpool University Press, pp. 227–279.

8 Semba (2012).

9 Bloch, C. E. (1917) Lidelser hos smaabørn opstaaet paa grund af fedtmangel: xerophthalmia et dystrophia alipogenetica. Ugeskrift for Læger 79, 309–325, 349–370; Blegvad, O. (1923) Om xerophthalmien og dens forekomst i Danmark i aarene 1909–1920. Copenhagen, Aalborg Stiftsbogtrykkeri.

10 Bloch (1917); Plum, P. (1957) C. E. Bloch. In memorium. Acta Pædiatrica 41, 505–509.

11 Monrad, S. (1917) Alimentær gastro-enteritis og xerophthalmi i barnealderen. Ugeskift for Læger 79, 1177–1196.

12 Bloch, C. E. (1921) Clinical investigation of xerophthalmia and dystrophy in infants and young children (xerophthalmia et dystrophia alipogenetica). Journal of Hygiene 19, 283–304.

13 Breast milk is a rich source of vitamin A.

14 Davis, L. E., Engerman, S. L. (2006) Naval blockades in peace and war: an economic history since 1750. New York, Cambridge University Press, p. 163.

15 Blegvad, O. (1924) Xerophthalmia, keratomalacia and xerosis conjunctivae. American Journal of Ophthalmology 7, 89–117.

16 Bloch, C. E. (1924) Blindness and other diseases in children arising from deficient nutrition (lack of fat-soluble A factor). American Journal of Diseases of Children 27, 139–148.

17 Bloch, C. E. (1919) Klinische Untersuchungen über Dystrophie und Xerophthalmie bei jungen Kindern. Jahrbuch für Kinderheilkunde und physische Erziehung 89, 405–441.

18 Anon (1924) Why xerophthalmia deserves attention. Journal of the American Medical Association 83, 1048–1049.

19 'Miles' [Maurice, J. F.] (1902) Where to get men. Contemporary Review 81, 78–86.

20 Bryant, L. S. (1913) School feeding: its history and practice at home and abroad. Philadelphia, J. B. Lippincott Company, p. 28.

21 Great Britain (1904) Report of the Inter-Departmental Committee on Physical Deterioration. Vol. I. – Report and appendix. London, His Majesty's Stationery Office, p. 57.

22 Anon (1905) Medical inspection and feeding of school children. British Medical Journal 2, 1409–1410.

23 Great Britain Parliament (1906) Education (Provision of Meals) Act 1906. London, H. M. Stationery Office.

24 Great Britain Parliament (1914) Education (Provision of Meals) Act 1914. London, H. M. Stationery Office.

25 Rosenau, M. J. (1912) The milk question. Boston, Houghton Mifflin Company.

26 McKee, F. (1997) The popularization of milk as a beverage during the 1930s. In Smith, D. F. (ed.) Nutrition in Britain: science, scientists and politics in the twentieth century. London, Routledge, pp. 123–141.

27 FD 1/3790. Accessory Food Factors Committee. Minutes of meeting, May 31, 1921.

28 Corry Mann, H. C. (1926) Diets for boys during the school age. Medical Research Council Special Report Series No. 105. London, His Majesty's Stationery Office.

29 de Onis, M. (2008) Child growth and development. In Semba, R.D., Bloem, M.W. (eds) Nutrition and Health in Developing Countries. Second edition. Totowa, New Jersey, Humana Press, pp. 113-138.

30 McCollum, E. V., Parsons, H. T., Kalmbach, E. (1924) The nutritional value of milk. Rogers, L. A., Lenoir, R. D. (eds) World's Dairy Congress, Washington, D.C., 2–10 October 1923. Washington, US Government Printing Office, pp. 421–437.

31 Vitamin D deficiency was unlikely to be a problem among these boys due to a great deal of direct sunlight exposure.

32 FD 1/3790. Corry Mann, H. C. Letter to W. Fletcher, June 20, 1921.

33 FD 1/3790. Corry Mann, H. C. Letter to A. W. J. MacFadden, January 22, 1922.

34 Food and Nutrition Board, Institute of Medicine (2001) Dietary Reference Intakes for Vitamin A, Vitamin K, Arsenic, Boron, Chromium, Copper, Iodine, Iron, Manganese, Molybdenum, Nickel, Silicon, Vanadium, and Zinc. Washington, D.C., National Academy Press.

35 One cup of watercress=34 grams in the US Department of Agriculture nutrient database [Available at http://nbd.nal.usda.gov/ndb/foods/list].

36 FD 1/3790. Fletcher, W. Letter to H. C. Corry Mann, March 20, 1923.

37 FD 1/3790. Corry Mann, H. C. Summary of diet experiments: report. May 14, 1923.

38 FD 1/3790. MacFadden, W. A. J. Letter to W. Fletcher, October 19, 1923.

39 FD 1/3790. Corry Mann, H. C. Letter to W. Fletcher, January 16, 1924.

40 FD 1/3790. Fletcher, W. Letter to A. W. J. MacFadden, Feburary 4, 1924.

41 FD 1/3790. Fletcher, W. Nutrition studies at Woodford (35) by Dr. Corry Mann. Confidential report to members of the Medical Research Council, April 4, 1924. (The report includes the opinion given by F. G. Hopkins, as quoted here.)

42 FD 1/3791. Fletcher, W. Letter to D. N. Paton, June 24, 1924.

43 FD 1/3791. Fletcher, W. Letter to D. N. Paton, May 15, 1924.

44 FD 1/3791. Paton, D. N. Letter to W. Fletcher, June 24, 1924; FD 1/3791. Memorandum on a report by Dr. Corry Mann on work carried out at Woodford, July 8, 1924.

45 FD 1/3791. Fletcher, W. Letter to D. N. Paton, June 27, 1924.

46 FD 1/3791. Fletcher, W. Memorandum on Nutrition Studies at Woodford, July 15, 1924.

47 FD 1/3791. Corry Mann, H. C. Letter to W. Fletcher, October 14, 1924.

48 FD 1/3791. Chick, H. Letter to W. Fletcher, March 12, 1926.

49 FD 1/3791. Corry Mann, H. C. Letter to A. L. Thomson, June 12, 1926.

50 FD 1/3792. Abbott, J. S. Letter to the Medical Research Council, February 19, 1927.

51 FD 1/3792. Millard, C. K. Letter to W. Fletcher, November 23, 1927.

52 FD 1/3792. Thomson, A. L. Letter to the Secretary, the National Milk Publicity Board, November 25, 1927.

53 FD 1/3792. National Milk Publicity Council, Inc. Letter to A. L. Thomson, November 27, 1927.

54 Orr, J. B. (1928) Influence of amount of milk consumption on the rate of growth of school children. Preliminary report. British Medical Journal 1, 140–141.

55 Leighton, G., Clark, M. L. (1929) Milk consumption and the growth of school children. Second preliminary report on tests to the Scottish Board of Health. British Medical Journal i, 23–25.

56 Leighton, G., McKinlay, P. L. (1930) Milk consumption and the growth of schoolchildren. Edinburgh and London: His Majesty's Stationery Office.

57 Student (1931) The Lanarkshire milk experiment. Biometrika 23, 398–406.

58 Gosset's assessment of the Lanarkshire study foreshadowed the inclusion during the 1930s of statisticians in the design of large experiments involving humans.

59 Atkins, P. J. (2005) Fattening children or fattening farmers? School milk in Britain, 1921–1941. Economic History Review 58, 57–78; Atkins, P. J. (2007) School milk in Britain, 1900–1934. Journal of Policy History 19, 395–427.

60 Wiley, A. S. (2005) Does milk make children grow? Relationships between milk consumption and height in NHANES 1999–2002. American Journal of Human Biology 17, 425–441; Wiley, A. S. (2009) Consumption of milk, but not other dairy products, is associated with height among US preschool children in NHANES 1999–2002. Annals of Human Biology 36, 125–138.

61 Hoppe, C., Mølgaard, C., Michaelsen, K. F. (2006) Cow's milk and linear growth in industrialized and developing countries. Annual Review of Nutrition 26, 131–173.

62 Gauger, M. E. (1936) Vitamins and your health. Fourth edition. New York, Robert M. McBridge & Company, pp. 27–28.

63 Great Britain. Board of Education. Medical Department. (1928) The health of the school child. Annual report of the Chief Medical Officer of the Board of Education for the year 1927. London, His Majesty's Stationery Office.

64 Great Britain. Board of Education. Medical Department. (1933) The health of the school child. Annual report of the Chief Medical Officer of the Board of Education for the year 1932. London, His Majesty's Stationery Office, p. 138.

65 Great Britain. Board of Education. Medical Department. (1947) The health of the school child. Annual report of the Chief Medical Officer of the Board of Education for the years 1939–1945. London, His Majesty's Stationery Office, p. 23.

66 Council on Foods and Nutrition, American Medical Association. Fortification of foods with vitamins and minerals. J Am Med Assoc 1939;113:681.

67 Semba, R. D. (2007) The impact of improved nutrition on disease prevention. In Ward, J. W., Warren, C. (eds) Silent victories: the history of practice of public health in twentieth-century America. New York: Oxford University Press, pp.163–192.

68 Bauernfeind, J. C., Arroyave, G. (1986) Control of vitamin A deficiency by the nutrification of food approach. In Bauernfeind J. C. (ed.) Vitamin A deficiency and its control. New York, Academic Press, pp. 359–388.

69 Gunderson, G. W. (1971) National School Lunch Program background and development. Food and Nutrition Service – 63. Washington, D.C., Food and Nutrition Service, US Department of Agriculture.

70 von Graefe, A. (1866) Hornhautverschwärung bei infantiler Encephalitis. Albrecht von Graefe's Archiv für Ophthalmologie 12, 250–256.

71 Hirschberg, J. (1868) Über die durch Encephalitis bedingte Hornhautverschwärung bei kleinen Kindern. Berliner klinische Wochenschrift 5, 324–326.

72 Stephenson, S. (1910) On sloughing corneæ in infants: an account based upon the records of thirty-one cases. Ophthalmoscope 8, 782–818.

73 Spence, J. C. (1931) A clinical study of nutritional xerophthalmia and night-blindness. Archives of Disease in Childhood 6, 17–26.

74 Blackfan, K. D., Wolbach, S. B. (1933) Vitamin A deficiency in infants. A clinical and pathological study. Journal of Pediatrics 3, 679–706.

75 Widmark, E. (1924) Vitamin-A deficiency in Denmark and its results. Lancet i, 1206–1209.

76 Cramer, W. (1924) An address on vitamins and the borderland between health and disease. Lancet i, 633–640.

77 Cowell, S. J. (1932) The principles of nutrition in preventive medicine. Royal Sanitary Institute Journal 53, 242–247.

Chapter 7

Rise of the 'Anti-Infective Vitamin'

Hidden hunger is starving millions. Are you a victim?
D. T. Quigley (1943), *The National Malnutrition*

The rickets studies of pharmacologist Edward Mellanby had earned him worldwide renown, and by 1925, he was conducting further rickets experiments to determine the properties of vitamins A and D. In his laboratory at the University of Sheffield, three hundred young dogs were being fed diets that included cod liver oil or butter, or cod liver oil that had been aerated at high temperatures, which destroyed its vitamin A but not its vitamin D [1]. One day Mellanby was summoned to come quickly to his lab: his dogs were dying. Fatal bronchopneumonia was sweeping through the animal quarters of the university's labs. The epidemic dealt a blow to Mellanby's ongoing study, but, fortuitously, the post-mortem examinations enabled him to make some observations of great interest. The dogs that had received the vitamin A-deficient diet – that is, the aerated and heated cod liver oil – had been susceptible to bronchopneumonia, while those on diets containing butter or cod liver oil that was still vitamin A-rich were unaffected by the disease [2]. Not a single case of bronchopneumonia was found in the dogs fed butter or unaerated cod liver oil [3].

The work of Osborne and Mendel and other biochemists had shown a relationship between vitamin A and infections in rats [4]. On the basis of these earlier findings and his own observations, Mellanby now speculated that the vitamin A-deprived dogs' susceptibility to infection might also have relevance to humans with respiratory illness or other infections. This possibility had to be investigated. Now, though, Mellanby would use rats instead of dogs, because rats were more economical and allowed for repeated experimentation to verify results.

Abating Childbed Fever: A Path with Forks and Obstacles

With support from the Medical Research Council, Mellanby hired Harry Norman Green (1902–1967), a recent graduate in medicine, to work with him on the relationship of vitamin A to susceptibility to infection [5]. Green was considering various positions at the time but decided to work with Mellanby because Sheffield's pharmacology department emphasized the experimental method, an approach to research that Green felt 'brought a bigger proportion of successes' [6]. For their first collaboration, Mellanby and Green conducted experiments with several hundred rats. Feeding their subjects experimental diets, they found, as they predicted, not only that xerophthalmia occurred in animals on diets deficient in vitamin A, but also that the same animals had weakened immunity and were susceptible to infection. They published their results in a paper in the *British Medical Journal*, which also attempted to correct a misperception of vitamin A's effects:

> Since the recognition of vitamin D (the antirachitic vitamin) as an entity distinct from vitamin A, those with experience of nutritional work have felt that to call vitamin A the 'growth-promoting' vitamin is a misnomer, for good growth often takes place in its absence if the diet is otherwise complete. In fact, when growth ceases owing to the single absence of vitamin A from the diet, it often means that the animal is definitely ill – in the sense. . . of having developed some, and often a widespread, infective condition. . . .. (I)t is, in fact, difficult to avoid the conclusion that an important, and probably the chief, function of vitamin A from a practical standpoint is an anti-infective agent [7].

To this they added, '. . .an extensive experience of nutritional work suggests that vitamin A is more directly related to resistance to infection than any other food factor of which we are aware'.

Mellanby and Green decided to look into a possible application of these findings to infections affecting humans. Many new mothers, for example, suffered from childbed fever, a common bacterial infection that affected women who had just given birth (the medical term, puerperal sepsis, comes from the Latin *puer* meaning child and *parere*, to give birth; the puerperium is the period immediately after delivery (textbox 7–1). Well into the 1930s, puerperal sepsis was the leading cause of childbirth deaths [8]. The most common puerperal infection was caused by the streptococcus bacterium; streptococcal infection could spread from the genital tract, where it originated, to the abdominal cavity, from which it could travel in the bloodstream throughout the body. Moreover, the disease could pass from to patient to patient via medical personnel who failed to take proper hygienic measures. With antibiotics (specifically, sulfa drugs, discussed below) not yet extant, treatments to improve the survival prospects of women in childbirth were much needed. Green and Mellanby's experiments suggested that vitamin A deficiency might well play a causal role in puerperal sepsis [9].

Textbox 7–1. Puerperal sepsis, contagion, and hygiene

Until the recognition of microbes and disease and the importance of hygiene in the late nineteenth century, puerperal sepsis was a major cause of death among women and was widely attributed to mysterious 'miasmas' in the air. Even before the validation of the germ theory of disease, progress against this puzzling killer came with the observations of Hungarian obstetrician Ignaz Semmelweis (1818–1865). Semmelweis noted that routine hand washing by doctors and midwives could substantially reduce mortality from puerperal sepsis [10]. Although the main cause of puerperal sepsis, beta-hemolytic streptococci bacteria, had not yet been identified, hand washing was observed to reduce the chance that a doctor, midwife, or nurse involved in deliveries could transmit the bacteria from infected women to uninfected ones.

With better hygienic practices, death rates began to fall from puerperal sepsis in England and Wales from about six per one thousand births by the late nineteenth century to just above one per one thousand in the 1920s [11]. By 1925, it became clear that beta-hemolytic streptococci were an important cause of puerperal fever, and thorough hand washing, soaking instruments in antiseptic solutions, and application of topical antiseptics to the mothers' external genitals were all thought to help reduce transmission of the bacteria among childbearing women.

Vitamin A made a brief appearance as a promising therapy for the prevention of complications related to puerperal sepsis, but the results were quickly eclipsed by Leonard Colebrook's demonstration that prophylactic use of sulphonamide drugs could greatly reduce the incidence of puerperal sepsis.

In general, pregnancy exacerbates a woman's susceptibility to the effects of vitamin A deficiency because of increased demand for the vitamin not only from her own body but also from the developing embryo/fetus. Mellanby and Green therefore proposed to conduct a study to determine whether vitamin A supplementation could reduce the occurrence and/or severity of puerperal sepsis. Sir Walter Fletcher, director of the Medical Research Council, which had supported Corry Mann's milk studies (chapter 6), informed Mellanby that the council was enthusiastic about Mellanby's plan, but he had trepidations [12]. 'Your scheme about pregnant women frightened us, of course. Your phrase "the proper feeding of 500 pregnant women" suggests enormous expenditure, which we not only could not undertake but in propriety should not undertake. The Council nevertheless would greatly wish to help work of the kind, and are completely sympathetic.' Fletcher's reaction gave the impression that the council's experience with Corry Mann's milk experiments with boys was still fresh in his mind: in Corry Mann's trial, many subjects fared far better than they would have if not made to live on experimental diets. Mellanby and the council finally agreed on a study design that involved, for a pool of subjects, the pregnant women who came either to Sheffield's antenatal clinic of Jessop Hospital or Nether Edge Municipal Hospital (fig. 7.1).

One of every two women would be given a bottle of vitamin A, which was prepared as a syrup. No specific dietary instructions were given, but each woman given vitamin A was directed to take half a teaspoon of it daily until she went into labor. Most of the women took the vitamin A in the last trimester of pregnancy: some for at least a month, others took it for about two weeks. Mellanby noted [13] that, 'Vitamin A therapy was stopped on admission to the hospital (for delivery) and the staff of the hospital who were attending to the welfare of the patients had no knowledge as to which of them had had the prophylactic preparation of vitamin A. After discharge

Fig. 7.1. a Jessop Hospital in Sheffield, England, **b** Nether Edge Municipal Hospital. The two hospitals where Green and Mellanby conducted the vitamin A trial for pregnant women are still standing. Photograph of Nether Edge Hospital © Barbara A. West, with permission of English Heritage Images of England.

from the hospital, the notes were collected and records made of any sepsis which developed in the puerperium.'

The outcome measure of the study, of course, was incidence of puerperal sepsis – or at least that was the investigators' original plan, and, indeed, the percentage of women who developed the infection, according to British Medical Association criteria [13], was 1.09% in the vitamin group and 4.73% in the control group. From these data, a reasonable inference was that vitamin A could reduce the occurrence of serious cases of puerperal sepsis by more than 75%, as reported in the *British Medical Journal* [14]. But the study results extended beyond puerperal sepsis. The subjects also included women with fevers associated with bladder and breast infections (cystitis and mastitis), so it actually showed that the vitamin treatment reduced the occurrence of other types of post-partum infections as well [15].

The editors of the *Lancet* noted a fourfold difference in the disease rate between the groups taking and not taking the experimental vitamin preparation – so great a difference that it was 'unlikely to have occurred merely by chance'. The observation had potential implications for thousands of women. For example, each year, England and Wales together had some 600,000 births, and if the vitamin A treatment were widely applied, with 'a reduction of this puerperal sepsis morbidity-rate from 4.73% to 1.09%,' it could conceivably result in '. . .a reduction from 28,000 to 6,000 cases' [16].

But the dramatic results of the study did not lead Mellanby to advocate vitamin A supplementation for all pregnant women. Thinking conservatively, he felt that his results only signified [14] that 'an adequate supply of vitamin A must be given to the pregnant woman. It should be the aim of all concerned with ante-natal welfare to see that the diet is rich in natural sources of vitamin A. Milk, egg yolks, green vegetables, carrots, and butter should be taken unsparingly. . .'.

Samuel J. Cameron, a gynecologist at the Scotland's County of Lanark Maternity Hospital, soon replicated Green and Mellanby's findings. Cameron had trained under Sir John Bland-Sutton, the physician who had cured rickety lion cubs in London by feeding them cod liver oil (chapter 5). In February 1932, Cameron read a paper before the Edinburgh Obstetrical Society in which he presented the results of his own investigations into the use of vitamin A in the prevention of puerperal sepsis.

At the time, a preventive called serum treatment was in use against puerperal sepsis. Its basis was serum taken from convalescing patients with streptococcal infection. Produced by pharmaceutical companies, this 'antistreptococcic serum' was expensive. In Cameron's trial, two-hundred thirty-five pregnant women were divided into four treatment groups: those given vitamin A, those given the serum treatment, those given both vitamin A and serum treatment, and as a control, those given no treatment. In the four treatment groups, the respective occurrence of puerperal fever was: 5%, 1.6%, 1.7%, and 14% [17]. The results of Cameron's trial closely resembled and added another aspect to Mellanby and Green's studies (table 7.1):

For a while, serum treatment upstaged vitamin A as a possible weapon against infectious diseases. In the 1930s, serum treatment was widely used against infections, because it was thought to help neutralize the infection, shorten the course of illness, and minimize complications [18]. The case for serum therapy originated in

Table 7.1. Two vitamin A/puerperal sepsis studies

Treatment for pregnant women	Percent incidence of puerperal sepsis
Mellanby and Green, 1931	
With vitamin A	1.09
Without vitamin A	4.73
Total percentage reduction in puerperal sepsis resulting from vitamin treatment	65
Cameron, 1932	
With vitamin A	5.0
With infection serum	1.6
With vitamin A and serum	1.7
Without vitamin A or serum	14.0
Total percentage reduction in puerperal sepsis resulting from vitamin treatment	75

1890, when Emil von Behring and Shisasaburo Kitasato found that mice inoculated with virulent tetanus bacilli could be protected from infection if immunized with blood drawn from rabbits immunized with virulent tetanus bacilli; mice not immunized with protective rabbit serum died when similarly inoculated. The experimenters concluded that something in the blood of the immunized rabbits had the ability to neutralize or destroy the toxin [19]. By the turn of the twentieth century, pharmaceutical companies and US government laboratories were producing serum by inoculating horses with tetanus or diphtheria and then collecting the serum for treatment of humans stricken with either of these diseases [20]. Serum therapy was also attempted for typhoid fever, pneumococcal pneumonia, and meningococcal meningitis. For diphtheria, it yielded some success, but results for tetanus and meningococcal meningitis proved less impressive. It became apparent that some bacteria, such a pneumococcus, had different types, and laboratories were needed to do the typing so that each patient received the appropriate serum.

After the vitamin A trials, Green conducted further investigations to measure the vitamin A reserves of women with active puerperal sepsis. Osborne and Mendel had shown in 1918 that the liver stores a reserve of vitamin A (chapter 5); now Green would measure the vitamin A stores in women's livers. To do this, he analyzed liver samples from female cadavers [21]. The livers of women who had died of puerperal sepsis and had received no supplementary vitamin A had the lowest vitamin A levels – about one-third the levels found in the livers of women whose deaths were caused by accidents [22]. Green's study appeared in the *Lancet* along with two other studies of vitamin A reserves in the human liver. The latter two showed that, in general, persons with chronic diseases [23] and severe widespread bacterial infections [24] had lower vitamin A concentrations in their livers than did accident victims (i.e. people whose deaths had not resulted from infections).

After Cameron's trial in Lanark confirming the findings of Green and Mellanby, one physician commented, '(The) results were so good that it might be worthwhile to consider their national application' [17]. But in the following years, vitamin A supplements were not widely adopted for prevention of puerperal sepsis, for two major reasons. First, as noted above, Green had advocated that pregnant women increase their dietary intake of vitamin A-rich foods rather than take vitamin A supplements, and Mellanby concurred. By this time, Mellanby had succeeded Sir Walter Fletcher as secretary of the Medical Research Council. This placed him advantageously to advocate improvement of the British diet and to foster further research on vitamin A as the 'anti-infective vitamin'.

The second reason was the discovery of the antibiotic sulfanilamide by Gerhard Domagk (1895–1964), a pathologist and bacteriologist at Germany's I.G. Farben Industrie, which produced dyes for textiles. When Domagk set out to develop a compound that could cure bacterial infections, his chemist colleagues provided him with azo dyes to test in mice [20]. One of the azo dyes (later named Protonsil) cured mice into which Domagk had injected lethal amounts of streptococcus bacteria; Domagk published this work in 1935 [25]. The following year, Leonard Colebrook at Queen Charlotte's Hospital in London showed that therapy with the same azo dye greatly reduced human mortality of puerperal sepsis (textbox 7.2) [26]. By the late 1930s, sulfonamide antibiotics were used widely for the treatment of puerperal sepsis and other infections, saving many lives.

Textbox 7–2. Antibiotics

The year after Gerhardt Domagk cured mice of streptococcus infection with azo dye, Leonard Colebrook, collaborating with Meave Kenny at Queen Charlotte's Hospital, treated 38 women with puerperal sepsis using sulfanilamide (Protonsil). The death rate among the treated women was 8%, compared with death rate of 26% of those not treated or before the start of treatment [26].

Beside I. G. Farben, other laboratories soon developed other sulfonamide antibiotics such as sulfapyridine, sulfadiazine, and sulfisoxazole, which soon superseded sulfanilamide for treatment of puerperal sepsis, pneumonia, and meningitis. Soon sulfonamide antibiotics were used to treat gonorrhea, a scourge of troops during World War II, but early success was followed by disappointment when sulfonamide-resistant gonorrhea began to appear among the troops [27].

The sulfa antibiotics were soon superseded by penicillin, which was originally discovered by Alexander Fleming in 1928. Fleming noticed that staphylococcus bacteria grown on culture plates were killed by a mold that had contaminated the plates [28]. Not until 1941, however, was penicillin produced in sufficient amounts to treat even one person. With the outbreak of World War II, the factories of the British pharmaceutical industry were being bombed, and large-scale production of penicillin began in the United States. Penicillin proved to be an effective cure for many common infectious diseases and was widely used to treat both soldiers and civilians during and after World War II.

As the first scientist to identify an antibiotic, Gerhardt Domagk was awarded the Nobel Prize in Medicine or Physiology 1939 but could not accept the award [29]. The Nazi regime forbade German nationals to receive the Swedish prizes after the German pacifist Carl von Ossietsky, who revealed that Germany was rebuilding its air force in violation of the Treaty of Versailles, was awarded the Nobel Peace Prize in 1935.

A Gentle Warrior Confronts a Children's Predator

Green and Mellanby's work on vitamin A as the 'anti-infective vitamin' caught the attention of Joseph Bramhall Ellison, a London pediatrician who was conducting research on measles, a highly contagious and potentially severe infectious disease caused not by a bacterium but by a virus. Its symptoms include a general skin rash, high fever, cough, and conjunctivitis, and it can be accompanied by complications of pneumonia, diarrhea, ear infections, and eye problems such as corneal ulceration, keratomalacia, and blindness. Outbreaks of measles were common in industrialized countries. During Ellison's time, measles epidemics would regularly sweep through London every two years [30]. But once infected, children who survived a bout with measles acquired a lifelong immunity against future attacks. Measles epidemics usually resulted when the proportion of infants and children who had not been previously exposed to measles reached a certain level that allowed measles to spread through the community. These factors accounted for the regularity of measles epidemics in London and elsewhere. Measles was the leading infectious cause of death in children in London in 1930 [31]. Of those who died with measles, more than 80% did so of complications from pneumonia. As a pediatrician, Ellison encountered thousands of measles cases, many of which ended in death.

Vitamin A deficiency was not uncommon in poor London families, nor was corneal ulceration and blindness resulting from the combination of vitamin A deficiency and measles in their children [32]. It was in this environment that Ellison originated the important idea of using vitamin A supplementation to treat children with severe measles.

In July 1931, Frederick Norton Kay Menzies, the Medical Officer of Health and School Medical Officer of London, reviewed the records of that city's 1929–1930 measles epidemic and the progress that had been made in the control and treatment of the disease. He noted the importance of nutrition to children with measles and those at risk of catching it. In 1931, Menzies wrote, 'It is. . . important to bear in mind that the prophylactic influence of foods rich in vitamin A both in preventing infectious diseases and protecting those already infected against severe complications is also of very great importance, more especially in regard to children' [32]. Menzies's timely pronouncement came just two months before the outbreak of the 1931–1932 measles epidemic.

At the Grove Fever Hospital, which treated patients with infectious diseases, many deaths were attributed to measles: according to Ellison's pathology studies, a full 50% (seventy-five infants and children of roughly one-hundred fifty child deaths). Ellison recorded the widespread damage that measles can do to epithelial surfaces of the respiratory tract [33], comparing it to the similar damage known to occur in vitamin A-deprived animals [34]. In a paper published in 1932 in the *British Medical Journal*, Ellison [35] wrote, 'Now a disease which attacks epithelial defenses and whose incidence is greatest in those members of the community who are most likely to be suffering from various grades of vitamin deficiency will probably prove the best medium for testing the therapeutic properties of vitamin concentrates. Measles appears to me to fulfil these criteria. In measles the brunt of the attack falls on epithelial structures'. Ellison wanted to determine whether vitamin A could moderate the degree of illness and avert death from measles. 'In this country the disease afflicts most heavily the children of the poorest classes, among whom the greatest mortality is observed during the first eighteen months after weaning. . . most of them are suffering from a lack of suitable fats in the diet'. Green and Mellanby, meanwhile, had published their findings on the use of vitamin A against puerperal sepsis in October 1931, just as the first cases of measles appeared in the anticipated epidemic of 1931–1932. The epidemic started in London's largely impoverished East End districts of Stepney and Poplar, spread westward to Paddington and Hammersmith, and then crossed the Thames and infected such urban areas as Lewisham, Woolwich, and Tooting. Soon the entire city was in its grip; measles eventually affected 55,545 London children (fig. 7.2).

Established by the London County Council (the former Metropolitan Asylums Board) and located in Tooting Grove south of the Thames, the Grove Fever Hospital, with five-hundred eighty-two beds (fig. 7.3), was one of twelve so-called fever hospitals in London specializing in infectious diseases. Other diseases treated in the fever hospitals included diphtheria and whooping cough – also infections that commonly occurred as epidemics. Combined, the twelve fever hospitals and another eleven facilities had a total of more than nine thousand beds [36]. These hospitals served a population of more than four million people in Inner London at the time.

Conducting a seven-month trial at the Grove Fever Hospital, where he served as an assistant medical officer, Ellison divided into two equal groups six hundred measles patients under age five. The arrangement of the wards at Grove Fever Hospital probably facilitated the trial; each ward had forty-four beds, and the wards were housed in separate buildings connected by covered walkways. The children with measles in Ellison's study were assigned to wards in which they received either the standard measles care or the standard care plus vitamin A. The no-vitamin-supplement group, i.e. the control group,

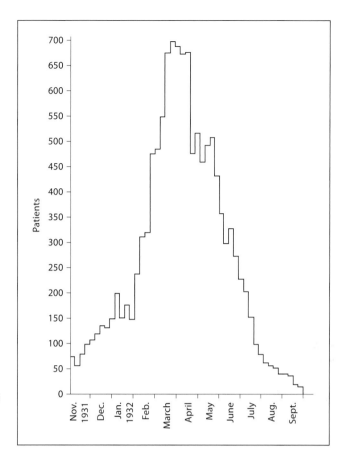

Fig. 7.2. Profile of the measles epidemic from November 1931 to September 1932 in London [31].

was fed the hospital's standard diet for children with measles. The vitamin preparation was a product called Adexolin, which was commercially produced from animal livers; its concentration of vitamin A was much higher than that in cod liver oil. Adexolin also contained vitamin D, as it is also found in animal livers. The hospital stays of these children ranged from seven days to three weeks, depending of the severity of their cases and whether or not they developed complications. Each child in the vitamin A treatment group was given about 20,000 IU of vitamin A per day [37]. The children's total oral vitamin A dose therefore ranged from approximately 140,000 to 420,000 IU [38].

Mindful that age is a significant factor in measles-caused death, Ellison subdivided the children by year of age when they were assigned to treatment to ensure that the age distribution was the same in each treatment group. By the end of the trial, eleven deaths had occurred in the vitamin A group and twenty-six in the control group. The findings showed that vitamin A had reduced mortality in all age groups except one: no deaths occurred among the children ages four and five (fig. 7.4). Most of the deaths had resulted from pneumonia as a measles complication, plus four from measles-related diarrhea. Of those four, three were in the control group; just one

Fig. 7.3. Grove Fever Hospital, Tooting, London. Courtesy St. George's Medical School Library.

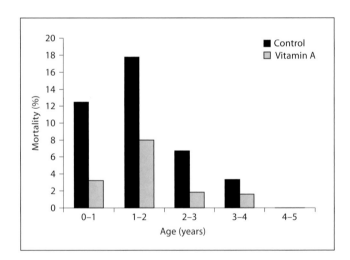

Fig. 7.4. Mortality among children in the control and vitamin A groups in Ellison's trial at Grove Fever Hospital, 1931–1932.

death occurred in the vitamin-supplement group. In statistical terms, Ellison's trial at Grove Fever Hospital showed that the vitamin A treatment reduced the children's measles deaths by 58%. Ellison attributed the reduction in deaths to the vitamin A alone, writing that, 'There are no grounds for supposing that the vitamin D in the concentrate exerted any specific beneficial effect on the course of the pneumonias'. (Although Ellison was the first scientist to show that vitamin A supplementation reduced measles-related deaths among children, later investigators overlooked his paper reporting the above findings – which drew worldwide attention when it was published. Comparable controlled trials of vitamin A in children with measles were not undertaken until the late-1980s (chapter 9) [38].)

After Ellison published his trial, the Grove Fever Hospital started giving vitamin A supplementation to children as part of routine care for measles. London's other fever hospitals, however, did not follow suit [39]. The Grove Fever Hospital adopted the vitamin A supplementation for measles patients partly out of collegiality, since Ellison was on the hospital staff and generally well-liked (textbox 7–3). By taking this maverick measure, Ellison's hospital undoubtedly save the lives of many young measles patients. During the epidemic of 1935–1936, the Grove Fever Hospital's total rate of measles complications among children was 18.8% – the lowest complication rate of all London's fever hospitals, in which the overall rate of measles complications was 34% [40]. Likewise, the death rate of children with measles at Grove was 3.4%, compared with 4.9% overall for the other fever hospitals. The death rate from measles reached 6.1% at Eastern Hospital and 6.4% at Brook Hospital [41]. (Vitamin A supplementation may account for the lower rate of complications and death in children with measles at the Grove Fever Hospital, but these findings are only suggested and not documented. Many factors – age, nutritional status, diagnostic criteria, and quality of nursing care – could have influenced these figures, but the historical record has no data either to corroborate or to disprove them.)

Textbox 7–3. An English eccentric who championed children

Joseph Bramhall Ellison was described as 'marked with that eccentricity of personal idiom that is peculiarly English' [42]. He was born in Kensington and attended Rugby School in Warwickshire, and then served in the Royal Engineers during World War I, where he suffered from severe shock after being 'buried alive' for several hours.

Ellison was graduated from Clare College, Cambridge University, and later took his MD in 1930. He entered the Fevers Hospital Service of the Metropolitan Asylums Board (which later became the London County Council) and worked at the Grove Fever Hospital, apparently content to work in a relatively junior position for twenty-five years, as 'the guiding motive of his professional life was his unfailing affection for the sick children under his care in the fever wards of the Grove'. As his friends later recalled, 'He would prefer to spend his Christmas in the hospital so that he could entertain the children and make them laugh, and there must be in the neighborhood of Tooting many hundreds of children who, with their parents, recall with deep affection his vivid and cheerful personality'.

> His colleagues considered Ellison brilliant. He was fluent in French and German, as well as having a remarkable knowledge of art, literature, and music. Indeed, the range of his interests was so broad that, as the writer of his obituary noted in 1953, '(I)t probably made it difficult for him to concentrate on any one course in life. . . (He) would take more pride in a poem he had written than in a paper published in one of the medical journals' [42].

Perhaps because he was a generally modest man, Ellison did not take up vitamin A supplementation as a cause as his career progressed despite the strong results from his trial at Grove Fever Hospital. His final contribution in vitamin A research was a study he conducted in collaboration with a colleague at Cambridge University. In this study, he showed that liver stores of vitamin A were lowest in young infants and in children who died with conditions such as pneumonia, widespread bacterial infection, and meningitis [43].

Ellison's trial of vitamin A therapy for measles was being conducted at a time when other investigators were keen on seeing whether serum therapy would work for measles. The serum used to treat measles was not produced in animals, but was rather collected from the parents of the child who had measles. Since the parents of children with measles were usually immune to the disease, the idea was that the parent's serum could neutralize or protect against the harmful effects of measles. Thirty milliliters of blood was obtained from the parents and then injected into the right and left buttocks of the child with measles [44]. Alternatively, serum could be collected from adults who were recovering from a measles attack, but measles was relatively rare in adults.

The use of convalescent and adult immune serum was adopted for measles in clinical practice in the 1930s. The number of people who received it was relatively limited: only about 12–15% of children admitted to England's fever hospitals with measles in the 1933–1934 and 1935–1936 measles epidemics were treated with serum therapy [45]. Serum therapy for measles was eventually abandoned by the late 1930s. (The measles virus was finally isolated in the 1950s, and a vaccine was developed introduced into wide use in the 1960s.)

A Vitamin's Short Stay at the Limelight

While scientists and lay writers impressed on a concerned public the necessity of vitamins to promote healthy growth and stave off such diseases as xerophthalmia, beriberi, scurvy, rickets, and pellagra, subclinical illnesses too eventually captured widespread attention. In *Your Meals and Your Money,* published in 1934, Gove Hambidge advocated a liberal supply of vitamin A in the diet to prevent subclinical deficiency [46]:

The mucous membranes throughout my body may be weakened through lack of *sufficient* vitamin A, though I never show a sign of xerophthalmia. . .. A severe shortage of vitamin A has far-reaching effects throughout the body, particularly in all mucous membranes. Now mucous membranes are the advance guards of the body against many kinds of bacterial infection; they protect eyes, mouth, nose, lungs, reproductive organs; and all in all, it would seem that vigorously healthy mucous membranes are as vital to general health as any other one thing we might name.

Although Hambidge did not explicitly say so, the efficacy of vitamin A in defending against childbed fever and measles was thought to be due to the nutrient's ability to protect the mucous membranes.

Other writers jumped in with their special diets. An example was the 'American Pep Diet' of one Dr. Inches, which was meant to combat all the problems that arise with subclinical vitamin deficiencies. The author of *The Forces in Foods* (1936), Dr. Inches was described as a cook, Broadway producer, dance director, pianist, explorer, and osteopathic doctor who, during his peripatetic career, 'consulted with the discoverer of vitamins, Sir F. Gowland Hopkins. . .'. The author's lament:

Men are more careful of their automobiles than they are of themselves. . .. Not for a minute would they allow a gasoline station attendant to pour crude oil into their car's gasoline tank, nor would they allow water to be poured in where the essential motor oil should go. Yet they sit down at a common cafeteria, lunchroom, or these roadside dining-cars and hot-dog stands and stuff their intestines full of the most stinking junk. . .' [47].

Dr. Inches advocated a diet full of such items as spinach, carrots, watercress, cantaloupe, cheese, egg yolk, mango, and sweet potatoes – all foods rich in vitamin A and recommended to build up resistance to respiratory infections and diarrhea.

Once vitamin A supplementation had shown promise as treatment for puerperal sepsis and measles, other studies of the 'anti-infective vitamin' soon followed. They involved school children, university undergraduates, and even industrial employees. By the 1940s, at least sixteen trials involving more than nine thousand subjects were in progress to determine whether vitamin A supplementation could lower the incidence of respiratory infections. With additional trials, investigators tried to determine the efficacy of vitamin A supplementation against pneumonia, scarlet fever, typhoid fever, and tuberculosis [48]. While some studies suggested that vitamin A supplementation affected the incidence or severity of disease, others produced no effect. Some trials were conducted in populations of adults whose dietary vitamin A intake was already adequate. Certain investigators expected to find that vitamin A supplementation would prevent or lessen the chance of a person's coming down with a common cold, pneumonia, or other infections. (In fact, vitamin A supplementation primarily benefits persons who have either clinically apparent vitamin A deficiency or subclinical vitamin A deficiency, i.e. low liver stores of vitamin A but no ocular signs of vitamin A deficiency. Vitamin A supplementation generally reduces the severity of certain infections but not the risk of contracting the infection. These now-proven certainties make sense of the mixed results of the thirty vitamin A supplementation trials carried out by 1940.)

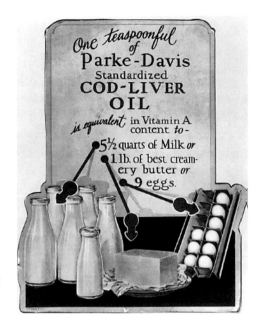

Fig. 7.5. Folding counter display advertisement for Parke-Davis cod-liver oil, circa 1935 (Collection of the author).

Daniel Thomas Quigley, a physician in Nebraska, alerted the public to *The National Malnutrition* (1943): 'Why we must have vitamins'. Vitamin deficiencies were widespread, he warned [49]: 'Hidden hunger is starving millions. . . Are you a victim?' Notwithstanding the mixed results of trials conducted in the 1920s and 1930s, vitamin A was widely taken up by the pharmaceutical companies, which advertised vitamin A as the 'anti-infective vitamin'. Companies urged vitamins on the public in general, especially on young mothers and pregnant women, exploiting a blend of fear, hope, guilt, and a modern image of motherhood [50]. Vitamin A-rich cod liver oil especially made for impressive advertisement (fig. 7.5), and the pharmaceutical company Squibb used the recent trials to advertise its product: 'Whooping cough, measles, mumps, chicken pox, scarlet fever may do greater harm than most mothers think. . .. (With Squibb cod liver oil) the children will have lighter cases, they recover quicker and (will be) . . . less likely to be left with some permanent injury, if they build up good general resistance in advance to fight them. . .. (O)ne precaution to build up the resistance of children. . . (is to give them) 'resistance-building' vitamin A. Vitamin A is the important factor which increases their fighting power in time of illness' [48]. Another ad read, 'A certain famous American doctor, whose life is being devoted to the study of malnutrition in children. . . tells mothers that one out of every three children in the United States is malnourished'. Still another raised mothers' self-doubt: 'Your family vitamin-starved? Impossible! Yet science now finds the average American family diet lacking in at least 3 important vitamins!' Industry was running a crash course in vitamins, and part of the scientific community helped.

With vitamin A deficiency recognized as a significant killer of children, especially in poor families, British physicians and legislators alike urged the vital nutrient on children. Cod liver oil was a cheap alternative source of vitamin A for the poor [51]: 'It is otherwise with the protective foods – milk, butter, eggs, fruit, and vegetables – which are the first to be sacrificed in time of hardship'. Cod liver oil therefore took a place in the morning routine of millions of children in Europe and the United States – with the approval of physicians and to the satisfaction of producers and vendors. Especially for children in poor families, cod liver oil became the primary – if hardly enjoyed – source of vitamin A.

Commercial fisheries in Newfoundland, New England, and Norway produced much of the world's supply of cod liver oil – hence, the world's vitamins A and D. The United States produced and imported a total of 640,000 gallons of cod liver oil in 1929 [52]. In the early 1930s, England's annual cod liver oil consumption reached 500,000 gallons [53]. As the *British Medical Journal* noted, 'cod-liver oil was in use in almost every working-class household, and local authorities spent considerable sums in purchasing bulk supplies for hospitals and sanitoriums' [54].

Protests arose in 1932 in Great Britain's House of Commons over a proposed tax on cod liver oil. Child mortality would increase, the bill's opponents argued, if the price of cod liver oil became too expensive for poor families. The politician William McKeag, who supported an amendment to exempt cod liver oil from the proposed taxation, noted that many a child in the north of England 'owed its life to being able to obtain cod-liver oil' [53]. Harriette Chick lent scientific weight to McKeag's position, writing in the *Lancet*, 'It is evident that any steps which may raise the price and lower the consumption of cod-liver oil, especially in the winter, would have deleterious effect upon the health of the population, involving particularly the well-being of the children of the poorer classes' [55].

By the 1940s, the realization was widespread that illness and death from infectious diseases could be averted by adequate nutriture with vitamin A [56]. Two medical collaborators at New York's Montefiore Hospital summarized the situation in a 1941 publication: 'It has been thoroughly established, during the last decade, that a deficiency of vitamin A in the diet, not only leads to interference with normal growth and well-being, but lowers the natural resistance of the individuals to infection' [57]. Medical practices and public health organizations that focused on the eradication of vitamin A deficiency confirmed the general acceptance of this theory. Such far-reaching organizations as the League of Nations Health Committee [58], the Women's Foundation for Health [59], the Council of British Societies for Relief Abroad [60], and the Medical Research Council of Great Britain [61] all emphasized the importance of ensuring adequate vitamin A intake to promote resistance to infectious diseases. The textbooks in nutrition, medicine, physiology, ophthalmology, and public health in many countries underscored this concern, inculcating upcoming generations with the importance of vitamin A to resistance to infectious disease.

Gradually, however, the emphasis in health-related research and practice shifted away from prevention of infectious diseases to curing them. The weight that swung the balance was antibiotics – first the sulfa drugs (textbox 7–2, above), then penicillin, which, by the mid-1940s, were both in widespread use. By that time, penicillin was being used to fight many bacterial infections.

With this shift, research and development on antibiotics largely overshadowed further studies of vitamin A as the 'anti-infective vitamin'. Great Britain's Edward Mellanby had termed vitamin A the 'anti-infective vitamin' with a caution. 'We were aware of the drawbacks of giving a label of this kind', he wrote, 'because the word 'infection' covers several different types of pathological phenomena, but we also recognized that it had the advantage of attracting the attention of workers to this important subject' [62]. Mellanby also made a prediction that was borne out:

'My own belief is that the relation of nutrition to infection will ultimately prove to be a subject of great practical importance. . .'.

References

1   PP/MEL/C.33. Notebook 'Diets 1918'. Notebook 'Dog Diets', 1001–2000.

2   PP/MEL/D.16. Lecture notes, Boston, 1930 'Vitamin A and infection'.

3   Mellanby, E. (1926) On diet and disease. With special reference to the teeth, lungs, and pre-natal feeding. British Medical Journal i, 515–519.

4   Osborne & Mendel (1914).

5   PP/MEL/B.1. Fletcher, W. Letter to Edward Mellanby, February 7, 1927.

6   PP/MEL/B13.25. Green, H. N. Letter to E. Mellanby, undated.

7   Green, H. N., Mellanby, E. (1928). Vitamin A as an anti-infective agent. British Medical Journal ii, 691–696.

8   Thomas, M. (1930) The epidemiology, bacteriology, and treatment of puerperal sepsis. Report of the Medical Officer of Health City of Glasgow 1930.

9   Mellanby, H. N., Green, H. N. (1929) Vitamin A as an anti-infective agent. Its use in the treatment of puerperal septicaemia. British Medical Journal i, 984–986.

10 Semmelweis, I. P. (1860) A gyermekágyi láz fölötti véleménykülönbség köztem s az angol orvosok közt Orvosi hetilap 4, 849–851, 873–876, 889–893, 913–915.

11 Kerr, J. M. M., Johnstone, R. W., and Phillips, M. H. (1954) Historical review of British obstetrics and gynaecology, 1800–1950. Edinburgh and London: E & S Livingstone Ltd., p. 215.

12 PP/MEL/B.1. Fletcher, W. Letter to Edward Mellanby, February 7, 1927.

13 PP/MEL/A60. Mellanby E. Maternal mortality. Public Health Congress (1930), Wednesday, November 19, 1930. Paper No. 7.

14 Green, H. N., Pindar, D., Davis, G., Mellanby, E. (1931) Diet as a prophylactic agent against puerperal sepsis. British Medical Journal ii, 595–598.

15 Loudon, I. (2007) An early Medical Research Council controlled trial of vitamins for preventing infection. Journal of the Royal Society of Medicine 1000, 195–198.

16 Anon (1931) Puerperal infection and dietary deficiencies. Lancet 1931, ii, 805–806.

17 Cameron, S. J. (1931) An aid in the prevention of puerperal sepsis. Transactions of the Edinburgh Obstetrical Society 52, 93–103.

18 Gunn, W. (1935) The treatment of measles. British Medical Journal 1, 597–599.

19 von Behring, E., Kitasato, S. (1890) Über das Zustandekommen der Diphtherie-Immunität und der Tetanus-Immunität bei Thieren. Deutsche medizinische Wochenschrift 16, 1113–1114.

20 Dowling, H. F. (1977) Fighting infection: conquest of the twentieth century. Cambridge, Mass, Harvard University Press.

21 Green used a chemical reaction, the antimony trichloride color reaction, that was widely used at the time to measure vitamin A concentrations in blood and tissues.

22 Green, H. N. (1932) Vitamin-A content of the liver in puerperal sepsis. Lancet ii, 727–726.

23 Wolff, L. K. (1932) On the quantity of vitamin A present in the human liver. Lancet ii, 617–620.

24 Moore, T. (1932) Vitamin-A reserves of the human liver in health and disease with special reference to the scope of vitamin A as an anti-infective agent. Lancet ii, 669–974.

25 Domagk, G. (1935) Ein Beitrag zur Chemotherapie der bakteriellen Infektionen. Deutsche medizinische Wochenschrift 61, 250–253.

26 Colebrook, L., Kenny, M. (1936) Treatment of human puerperal infections, and of experimental infections in mice, with Protonsil. Lancet i, 1279–1286; Colebrook, L., Kenny, M. (1936) Treatment with Protonsil of puerperal infections due to haemolytic streptococci. Lancet i, 1319–1322.

27 Lesch, J. E. (2007) The first miracle drugs: how the sulfa antibiotics transformed medicine. New York, Oxford University Press.

28 Fleming, A. (1929) On the antibacterial action of cultures of a penicillium, with special reference to their use in the isolation of B. influenzae. British Journal of Experimental Pathology 10, 266–236.

29 Domagk later accepted the Nobel Prize in 1947.

30 London County Council (1938) Measles: report of the Medical Officer of Health and School Medical Officer on the measles epidemic, 1935–1936. London, London County Council, p. 5.

31 London County Council (1933) Measles: report of the Medical Officer of Health and School Medical Officer on the measles epidemic, 1931–1932. London, Central Public Health Committee.

32 Currie, M. R. (2005) Fever hospitals and fever nurses: a British social history of fever nursing: a national service. London, Routledge Taylor and Francis; London County Council (1931) Measles: Report of the Medical Officer of Health and School Medical Officer on the measles epidemic, 1929–30. London, London County Council, pp. 64–65.

33 Ellison, J. B. (1931) Pneumonia in measles. Archives of Disease in Childhood 6, 37–52.

34 Wolbach, S. B., Howe, P. R. (1925) Tissue changes following deprivation of fat soluble A vitamin. Journal of Experimental Medicine 42, 753–777.

35 Ellison, J. B. (1932) Intensive vitamin therapy in measles. British Medical Journal ii, 708–711.

36 Metropolitan Asylums Board (1927) Annual report for the year 1926–27. London, Harrison and Sons.

37 20,000 IU of vitamin A is equivalent to 6 mg of retinol.

38 Semba, R. D. (2003) On Joseph Bramhall Ellison's discovery that vitamin A reduces measles mortality. Nutrition 19, 390–394. The amount of vitamin A received can be compared with current WHO/UNICEF/IVACG recommendations to provide children with acute measles with 200,000 IU upon admission and 200,000 IU the following day.

39 London County Council (1938) Methods of hygiene, isolation, nursing, antibiotic (Protonsil) use, oxygen tent therapy, anti-toxic sera, drug treatment, and serum therapy of the fever hospitals are described in detail, with no mention of vitamin A therapy.

40 London County Council (1938), pp. 64–65; the complication rates at Eastern, North-Eastern, North-Western, Western, South-Eastern, Park, Brook, Joyce Green, and South-Western fever hospitals were 44.5%, 45.3%, 44.2%, 34.4%, 24.9%, 43.6%, 37.3%, 25.5%, and 37.3%, respectively.

41 London County Council (1938), pp. 60–61; mortality rates were calculated after excluding Joyce Green, which only admitted older patients, and Southern, which admitted few patients with measles.

42 Anon (1953) Obituary for Joseph Bramhall Ellison. British Medical Journal ii, 1107.

43 Ellison, J. B., Moore, T. (1937) Vitamin A and carotene. The vitamin A reserves of the human infant and child in health and disease. Biochemical Journal 31, 165–171.

44 London County Council (1931), p. 18.

45 London County Council (1938), p. 45.

46 Hambidge, G. (1934) Your meals and your money. New York, Whittlesey House, pp. 93, 96.

47 Inches, H. V. H. (1936) The forces in foods. Boston, American Dietetic Research Foundation, p. 89.

48 Semba, R. D. (1999) Vitamin A as 'anti-infective' therapy, 1920–1940. Journal of Nutrition 129, 783–791.

49 Quigley, D. T. (1943) The national malnutrition. Milwaukee, Lee Foundation for Nutritional Research.

50 Apple, R. D. (1996) Vitamania: vitamins in American culture. New Brunswick, New Jersey, Rutgers University Press.

51 Anon (1932) Taxation of cod-liver oil. Lancet 2, 850.

52 Prescott, S. C., Proctor, B. E. (1937) Food technology. New York, McGraw-Hill Book Company, Inc.

53 Anon (1932) Parliamentary Intelligence. Ottawa and cod-liver oil. Lancet ii, 978–979.

54 Anon (1932) Medical notes in Parliament. Ottawa agreements: cod-liver oil. British Medical Journal ii, 861.

55 Chick, H. (1932) Correspondence. Taxation of cod-liver oil. Lancet ii, 919.

56 Heilbron, I. M., Jones, W. E., Bacharach, A. L. (1944) The chemistry and physiology of vitamin A. Vitamins and Hormones 2, 155–213.

57 Perla, D., Marmorston, J. (1941) Natural resistance and clinical medicine. Boston, Little, Brown and Company.

58 Mixed Committee of the League of Nations (1937) The relation of nutrition to health, agriculture and economic policy. Document A.13.1937.II.A. Geneva, League of Nations, 61.

59 Rose, M. S. (1928) Choosing food for health. In: Handbook on positive health [no editor]. Revised edition. New York, Women's Foundation for Health, 1928, pp. 114–132.

60 Council of British Societies for Relief Abroad (1945) Nutrition and relief work: handbook for the guidance of relief workers. London, Oxford University Press.

61 Medical Research Council (1932) Vitamins: a survey of present knowledge. Special Report Series, No. 167. London, His Majesty's Stationery Office, 28–31.

62 Mellanby, E. (1934) Nutrition and disease: the interaction of clinical and experimental work. Edinburgh, Oliver and Boyd.

Chapter 8

Vitamin A Deficiency in Europe's Former Colonies

Some consider that the true public health weight of the problem. . . is obscured because its victims often die before they can be reported as blind.
C. Fritz, *Combating nutritional blindness in children: a case study in technical assistance in Indonesia* (New York, Pergamon Press, 1980)

The empire building and colonization activities of Western Europe's powerful nations during the sixteenth to the nineteenth centuries inevitably, if slowly, awakened the industrialized West to health conditions in impoverished parts of Asia and Africa. The Netherlands was one of the first of imperial powers to acknowledge and take some responsibility for people under its dominion affected by diseases that ran unchecked. As a secondary effect, this acknowledgment played a key role in establishing public health as a discipline in which field research and treatment continue be essential components and as an area of multinational concern.

Dutch Initiative versus the Free Market

The Laboratory for Pathological Anatomy and Bacteriology established by the Dutch government in the 1880s in Jakarta (former Batavia) undertook some of the earliest comprehensive field research on dietary deficiencies to be carried out in an imperial colony. Batavia is where Cornelius Pekelharing, then Christiaan Eikjman, Gerrit Grijns (chapter 5), and then Barend Jansen and Willem Donath, studied beriberi. The disabilities resulting from this neurological disorder, caused by a lack of dietary thiamin, affected thousands of indigenous people throughout Asia.

Even as beriberi was being brought under control, research efforts in nutrition focused on the study and control of vitamin A deficiency. European missionaries, physicians, and other travelers in the colonies brought home sporadic reports of xerophthalmia from Africa, Brazil, China, and the East and West Indies [1]. A French physician observed in 1900 that the people of Sumatra were much affected by *rondar manok* – night blindness' [2]. Nearly two decades later, W.A. Wille, a British ophthalmologist working in Semarang, described one hundred and fifty children ages five and younger who had xerophthalmia and keratomalacia; following the example of Carl Edvard Bloch in Copenhagen, Wille cured the children by giving them cod liver oil (chapter 6) [3]. An ophthalmologist named Louwerier working on a plantation in eastern Sumatra noted that night blindness was a common complaint among young Javanese laborers; in this instance, interestingly, the symptom disappeared if the workers' financial circumstances improved [4].

Early medical reports suggested that xerophthalmia and keratomalacia were common in the Dutch East Indies. Unlike many other colonists, however, the Dutch remained in the forefront of field research, and also, in care. Johannes Tijssen (1881–1948), a Dutch physician trained in Leiden, investigated the origins of blindness in Aceh on the western tip of Sumatra and concluded that vitamin A deficiency was the single greatest cause [5]. While working on an oil palm plantation for sixteen years, Tijssen established a medical service. In 1928, he returned to the Netherlands for long enough to study ophthalmology, which would augment his ability to serve patients in the Dutch East Indies. He worked in Sumatra, Java, and Borneo as an itinerant eye surgeon, conducting thousands of cataract operations. Tijssen reported that the rate of blindness among young boys was nearly thirty times higher in West Java than in the Netherlands. Nearly all this staggering rate of childhood blindness was attributable to vitamin A deficiency [6].

Expanded Dutch studies in rural villages in the late 1930s found vitamin A deficiency remained widespread among children [7]. Two institutes in Batavia, one named for the physician Christiaan Eijkman, and the other the Institute for Nutrition Research, undertook a series of dietary surveys after outbreaks of xerophthalmia and high mortality were reported among children in different areas of rural Java. Children's intake of vitamin A-rich foods, the studies found, was lower among boys than among girls. The discrepancy was ascribed to social custom: in many parts of Java, vegetables were deemed suitable fare for women and not men [8]. Overall, the Dutch investigators' findings largely corroborated the clinical observations of Masamichi Mori in Japan and others at the turn of the century (chapter 5): xerophthalmia and keratomalacia were associated with infections; the disease had a seasonal pattern; it was associated with poverty and a poor diet; and the chance of an affected child's dying was extremely high.

Dutch scientists were among the first to measure vitamin A levels in the blood of children, pregnant and nursing women, and other adults with xerophthalmia as well as in human breast milk [9]. They found children with xerophthalmia to have much lower vitamin A levels than children with healthy eyes [10]. Women living in poverty had lower levels of vitamin A in their breast milk than did women in better economic circumstances [11].

Reports of child blindness and deaths attributable to vitamin A deficiency in the Dutch East Indies had, in fact, started rising in the late 1920s [12]. Papers published in 1929 and 1931 reporting studies by ophthalmologist Sie Boen Lian and pediatrician Jacob Hijmans de Haas attest that hundreds of children in Java and Sumatra had keratomalacia [4, 13]. The two physicians described the lack of vitamin A in the diets

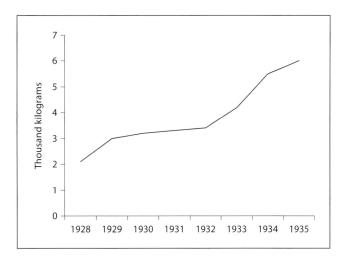

Fig. 8.1. Imports of sweetened condensed skim milk in the Dutch East Indies (1928–1935) [17].

of pregnant women, of women who were breastfeeding, of infants who were being weaned, and of young children. Matters only got worse, reaching what Haas termed the 'crisis period' of 1928–1935.

Ironically, the dairy industry was the cause. As early as the 1600s, Dutch colonists had introduced dairy practices in the Dutch East Indies. The industry there was limited and meant to meet the needs of the Dutch settlers, not indigenous people. Sometimes the commercial ships of the Dutch East India Company (*Vereenigde Oost-Indische Compagnie* or V.O.C.) carried cheese and butter from Holland, but because the voyage was extremely long and took ships from the chilly waters off Northern Europe into the tropical seas around the East Indies, most dairy products arrive spoiled and inedible. The innovation of canning enabled the shipment of tinned milk. By the 1880s, Switzerland's Nestlé Company and it competitor, the Anglo-Swiss Condensed Milk Company, were marketing their products in Africa and Asia – mainly to Europeans [14]. The two firms merged in 1905 under the name Nestlé. Five years late, Nestlé entered the market in Java.

The firm dispatched salesmen to rural villages, where they distributed samples of tinned, sweetened, condensed, skim milk – free. Even when a modest charge was introduced, this new product came with the advantage that it could be stored without refrigeration. Once opened, it did not spoil so easily as fresh milk because of its high sugar content [15]. So tinned milk was taken up by government and mission hospitals. For domestic use, it appeared in markets and villages shops and was hawked by street vendors. Nestlé kept the price low and the sugar content high, making it accessible to poor families and appealing to their children [16]. Soon lactating mothers took to replacing their own milk for their infants with Nestlé's. The importation of sweetened condensed skim milk tripled between 1928 and 1935 (fig. 8.1). Meanwhile, other firms also exploited the enthusiasm of Javanese mothers and their children for tinned condensed milk and entered the market in competition with Nestlé.

While children's consumption of tinned condensed milk rose, too did the rates of xerophthalmia and keratomalacia in the local eye hospitals. Suspecting a link, De Haas and his colleagues analyzed the vitamin A concentration in Java's five major tinned milks. Their findings: some milks had low concentrations of vitamin A, and sweetened skim milk had virtually none [17]. When tested in the laboratory on rats, animals fed sweetened skim milk developed xerophthalmia and died [18]. Even the tinned milk that contained some vitamin A lost its potency when mothers, trying to make the milk go farther, diluted it with water. In Sumatra, meanwhile, Louwerier found an especially high incidence of keratomalacia among adopted children, as 'the tinned milk becomes too expensive and is always diluted' [4].

De Haas called for a ban on the importation of sweetened skim milk to the Dutch East Indies, calling it a 'waste product' from Europe that resulted in blindness and death in infants and young children. For their part, the milk manufacturers dismissed the removal of vitamin A as a trivial matter. A Dutch scientist with the Friesland Condensed Milk Cooperative in Leeuwarden, G. S. de Kadt, vigorously defended skim milk as an ideal food for children, quoting official reports from the League of Nations:

> Milk, it is recommended, should represent a large proportion of the diet at every age. The London report approved the tendency displayed in certain countries to increase the consumption of milk to as much as one liter per day for pregnant women and nursing mothers, and to provide a comparatively large quantity for infants, children and adolescents. For these classes, it strongly recommends free or cheap milk distribution. *It calls attention to the nutritive value of skim milk, which, although it has lost its fat and fat soluble vitamins, retains many other valuable nutrients.* It therefore deplores the way in which the valuable food is wasted in many countries. . .. The Commission desires to draw attention to the high nutritive value of skimmed and separated milk, which although deprived of its vitamin A through removal of the fat, retains the protein, the B and C vitamins, the calcium and other mineral elements. The Commission deplores the large wastage in many countries of the valuable food [19].

Infants and young children fed sweetened skim milk continued to develop corneal ulceration and keratomalacia as well as diarrhea, depressed immunity, and increased susceptibility to infection. Many died. De Haas and others continued to advocate a ban on the importation of sweetened skim milk [20], and the debate reached the *Volksraad* (People's Council). De Haas and his colleagues showed that, during 1935–1939, xerophthalmia accounted for 5% of all pediatric admissions to the Central Civil Hospital in Batavia. Sweetened skim milk accounted for two-thirds of all cases of xerophthalmia of children under age two [21]. Indigenous people in colonial outposts elsewhere in Asia were similarly susceptible to the appeal of tinned milk.

Manufacturers in the industrialized West seized this opportunity. Alexander Dewar Williamson, director of the eye clinic at Singapore General Hospital, noted a gradual increase in the number of children with keratomalacia from 1935 onwards, and pointed out a paradox. The rising rate of the condition, he noted, '. . .was associated with a parallel rise in the standard of living. . .. (T)he working class Chinese mother, often a worker herself, found that she was increasingly able to buy milk, usually in the sweetened condensed form. This tendency was greatly fostered by the intensive propaganda of certain milk merchants. Most of the cases of keratomalacia found in this clinic have the history of almost exclusive feeding on sweetened condensed milk' [22]. Cicely Williams (1893–1992), a nutritionist and pioneer breastfeeding advocate, voiced outrage that milk companies were sending young women clad in doctor-like white coats to promote tinned milk in Singapore tenements. Williams articulated her outrage in 'Milk and Murder,' an address delivered in 1939 to the Singapore Rotary Club [23]. '(M)isguided propaganda on infant feeding should be punished as the most criminal form of sedition, and that those deaths should be regarded as murder' [24]. Finally, the outbreak of World War II silenced the rows over promoting sweetened condensed milk in the colonies, but the problem resurfaced (textbox 8–1).

Textbox 8–1. The public good takes on free enterprise

The Nestlé Company's marketing of substitutes for breast milk in developing countries drew heated objection in the 1970s. The product under criticism was infant formula. Sold in powder form, infant formula was advertised as a sure source of complete infant nourishment – once it was mixed with water. Until it was needed, the powder was inert in its sealed package, hence proof against spoilage and external contamination. And it was cheap.

What escaped notice at first was that the water needed to liquefy the powdered milk might itself be contaminated; a water supply that was free of contaminants, hence safe to drink without being boiled, was an asset mostly of the industrialized West. Even as onetime colonies gained independence in the course of the mid-twentieth century, safe water supplies did not come with it. Breastfeeding, meanwhile, was increasingly discouraged as old-fashioned, even primitive. So mothers, seeking to be modern in their way of life and taking advantage of the readily available and inexpensive formula preparations being aggressively marketed, found themselves inadvertently feeding their infants contaminated formula.

This time outrage among health professionals and the general public coalesced, and the marketing of substitutes for breast milk was met with a boycott. Although many companies, both European and American, were mining the rich vein of mothers and their infants in the developing world, the boycott focused on Nestlé products [25]. Nestlé accounted for the largest proportion of infant formula sales in developing countries. In May 1981, delegates to the World Health Organization entertained the proposal of an international code of conduct restricting the advertising and marketing of baby formula. Physicians, government officials, nongovernmental organizations in public health, women's rights groups, and conscience-driven individuals around the world who had supported the boycott of Nestlé now supported the WHO code.

Political conservatives in the United States, however, argued that the proposed WHO code ran counter to the ideals of free speech and free trade. Accordingly, President Ronald Reagan ordered the US representative to WHO to cast a No vote

Health in the Developing World Becomes a Multinational Concern

The end of the war brought increased attention to health problems in many developing countries. One result was the establishment in 1946 of the United Nations International Children's Emergency Fund (UNICEF). The World Health Organization (WHO) was created two years later.

Strangely enough in this postwar atmosphere of rising multinational sensitivity and cooperation, a military strategy of the 1950s brought about further delineation of the vitamin A deficiency problem in developing countries. During the Korean War, malnutrition found to be common among the troops of the Republic of Korea and the Republic of China (Taiwan) raised concerns among commanders on the US side that its troops, too, might become impaired by the same condition. The Interdepartmental Committee on Nutrition for National Defense was therefore created to spearhead an effort 'to deal with nutrition problems of technical, military, and economic importance in foreign countries in which the United States has a special interest' [27]. The committee first had nutrition surveys conducted among military personnel and then expanded them to civilians in various communities. Thousands of children in selected countries from Asia and Africa to the Caribbean and Latin America were examined for Bitot's spots and other manifestations of vitamin A deficiency, and blood samples were taken to measure vitamin A levels. Vitamin A deficiency was found to be a major health problem in many countries, including East Pakistan, Ethiopia, Jordan, Lebanon, Thailand, and Vietnam [28].

Five years after the armistice that marked the end of the Korean War, the U.S. National Institutes of Health (NIH) collaborated with WHO to hold one of the first major international conferences to focus on vitamin A deficiency. The topic of the meeting, which was held in Princeton, New Jersey, on beriberi, iodine deficiency, and vitamin A deficiency [29].

H.A.P.C. Oomen (1902–1986), a Dutch physician nicknamed 'Janus' who worked in the East Indies, served for many years in regional hospitals and as a plantation doctor. World War II had interrupted his work and that of many colleagues, when the Japanese took many Dutch physicians as prisoners. After the war, however, Oomen resumed his nutrition work, and when Indonesia became independent in 1949, the new minister of health named him Indonesia's representative to WHO. One of his main tasks was to deal with xerophthalmia. 'In every hospital in Djakarta attended by poor children', Oomen lamented in 1953, 'perforated eyes may be discovered daily. . .. (Xerophthalmia) is known to occur everywhere, with undefined variations according to season and locality. As is also true of many other diseases, the more one looks for xerophthalmia, the more one finds it' [30].

At the UNICEF conference, Oomen voiced his belief, based on his work in Indonesia, that young children with what was then considered 'mild' vitamin A deficiency (signaled by conjunctival xerosis and Bitot's spots) accounted for considerable proportion of child deaths in that part of the developing world. He regarded the months from twelve to twenty-four to be 'the worst (year) in the nutritional history of the Indonesian child. . . xerophthalmia creeps through the whole toddler period. . . and bacteria and viruses find him an easy and unprotected prey.' In the parts of Sumatra and Java where he worked, Oomen estimated that the prevalence of 'mild' xerophthalmia [31] to be 1–7% – '. . .enough to be responsible for a considerable part of toddler mortality' [32]. He expressed his rage publically, charging his colleagues in the international medical community with having not taken a single step toward solving the health problems of the developing world.

In a report on the global problem of vitamin A deficiency published in 1964, Oomen and two colleagues hypothesized a vicious cycle keeping vitamin A deficiency and infectious diseases insidiously linked. 'There appears to be a universal relation between infectious disease and xerophthalmia. . .' they wrote, continuing, 'Not only may deficiency of vitamin A itself play an important role in lowering the resistance to infection. . . but infectious diseases themselves predispose to and actually precipitate xerophthalmia' [33].

Vitamin A deficiency came under further international scrutiny at a 1963 conference held in northern Italy under the auspices of the 6th International Congress of Nutrition and titled 'How to Reach the Pre-School Child'. A consensus emerged that the time had come for emergency action – a crash program to awaken the world to the extent of child mortality and to take action on children's nutritional needs [34]. The National Academy of Sciences then held a follow-up international conference, 'Prevention of Malnutrition in the Pre-School Child', in 1964. The director of the NIH, William Sebrell, presided. The high death rates among young children in developing countries were finally receiving long-overdue attention: mortality rates for preschool children in developing countries could be forty times greater than in affluent countries. Malnutrition, including vitamin A deficiency, was cited as a significant reason for the higher mortality. The conference's final report carried the statements that, '(T)he mortality rate among malnourished children with xerophthalmia is very high. . .' and '. . .present evidence enforces the ominous conclusion that the incidence of xerophthalmia is increasing' [35].

Scientists at the meeting discussed pilot efforts for giving large doses of vitamin A to children at risk in countries where xerophthalmia was common [36], and some notable interventions followed. Paul György, a pediatrician at the University of Pennsylvania who had attended both the 1963 and 1964 conferences, reported on an intervention trial conducted in Indonesia: a teaspoon of vitamin A-rich red palm oil was given daily to preschool children in villages that had a high prevalence of vitamin A deficiency. After two months, the prevalence of xerophthalmia dropped from more than six percent to less than one percent. György was enthusiastic about the results, but practical. 'As stated in the recommendation of the [1963] conference, any 'crash action program' for the pre-school child should dovetail with long-term projects already in progress, such as maternal and child health centers, agricultural extension, community development and nutrition education' [37].

In India, meanwhile, Vulimiri Ramalingaswami (1921–2001), a distinguished physician and nutritionist who became director of the All India Institute of Medical Sciences, drew attention to the important association between diarrheal disease and vitamin A deficiency. The young children who appeared at the Nutrition Clinic of the Nutrition Research Laboratories in Coonor in southwest India sharpened his concern. He noted that diarrheal disease and abnormalities in the intestinal linings were consistent features of vitamin A deficiency both in experimental animals and in humans, and that the administration of vitamin A reduced the diarrheal disease in both [38]. In a therapeutic trial, Ramalingaswami showed that high doses of oral vitamin A could be used to treat diarrhea in children with xerophthalmia. 'From these observations', he noted, 'and from a consideration of the literature, it is concluded that a deficiency in vitamin A, diarrhea occurs which responds specifically to vitamin A' [39].

Ramalingaswami's trial could have made a turning point in knowledge about the effects of high doses of vitamin A on the severity of diarrhea in young children with vitamin A deficiency. But like Joseph B. Ellison's trial of high doses of vitamin A for vitamin A-deficient children with measles (chapter 7), the importance of the study went unrecognized for too long a time.

A group of scientists met in 1968 at the Pan American Health Organization, a regional office of WHO, was charged with studying the problem of vitamin A deficiency. Its report: 'From experiments in animals. . . it can be assumed that prolonged low intake of vitamin A and its precursors may have serious effects on growth and development and on resistance to infectious diseases'. The report went on, 'The regular occurrence of xerophthalmia cases in an area is indicative of a very serious preschool public health nutrition problem. The high case fatality rate of at least twenty-five percent contributes to underestimation of its magnitude' [40]. The group made six recommendations for locales where vitamin A deficiency was a problem:

- that infants and preschool children receive oral high-dose vitamin A supplements one to four times a year;
- that lactating women receive oral high-dose vitamin A immediately after delivery;

- that foods such as dried skim milk be fortified with vitamin A;
- that nutrition education be promoted;
- that agricultural production of vitamin A-rich foods be encouraged, and
- that professional training about vitamin A deficiency be fostered.

These recommendations were often repeated over the following years, with limited action. Into the 1970s, efforts to eliminate vitamin A deficiency in developing countries evolved with broad-based use of high-dose vitamin A supplements for children where vitamin A deficiency was endemic, and, specifically, with vitamin A fortification of sugar in Guatemala. By 1965, the Western Hemisphere Nutrition Congress had recognized the occurrence of high child morbidity and mortality attributable to vitamin A deficiency [41]. Central American and Panamanian scientists uncovered a widespread problem of vitamin A deficiency in the populations of their region: some 1.5 million children under age fifteen were suffering the effects of vitamin A deficiency [42]. Guillermo Arroyave and his colleagues at the Institute of Nutrition of Central America and Panama (INCAP) pioneered efforts to reduce vitamin A deficiency through food fortification (textbox 8–2).

Textbox 8–2. Vitamin A for food fortification

The biochemical form of vitamin A that is most commonly used for food fortification is retinyl palmitate (vitamin A palmitate). Chemically, it is the ester of retinol (vitamin A) and palmitic acid. This form of vitamin A is relatively stable and is well absorbed. Historically, the Swiss company Hoffman-La Roche was the world's only major supplier of retinyl palmitate, but starting in the late 1990s, the number of commercial laboratories worldwide that synthesize retinyl palmitate has grown tremendously. India alone has several private companies that manufacture vitamin A.

From a public health standpoint, for food fortification to be effective in reducing a population's micronutrient deficiency, the food to be fortified must be a dietary staple eaten daily with little or no variation. Further, the fortified food should reach the entire population. Of course, the fortification process must be economically feasible and have minimal effect on the cost of the food treated. The micronutrient with which the staple is treated must be chemically stable and undetectable by persons consuming it. Finally, to enable observation and measurement of results, location or processing and distribution must be finite and constant.

The population, conditions, and customs of Guatemala made this relatively small Latin American country a suitable place for testing dietary vitamin A fortification. A quite uniform diet based largely on maize suggested at first that maize flour was a candidate for fortification. But the fact that many families grew their own maize and dried and ground it themselves meant that fortification under controlled conditions by industry outside the household was not possible. Wheat flour was another possibility, but wheat products tended to be expensive and therefore used much less than maize flour, especially by poorer families. All families used salt, but salt was chemically not suitable for fortification with vitamin A [43]. Sugar, the scientists finally decided, was the most appropriate vehicle. Most Guatemalans consumed sugar, and in relatively consistent quantities across the population. In addition, it would simplify matters that all Guatemala's sugar came from just a handful of manufacturers.

The program to fortify sugar – necessarily a strong and mostly invisible effort – was presented to the authorities and the populace at large as a public health measure to protect children against illness, blindness, and death. But Guatemalan business executives, especially the sugar tycoons who would be bidden to shoulder the costs, balked at the program's likely expense. Succumbing to heavy lobbying, the Guatemalan Congress rejected a draft proposal for mandatory sugar fortification in September 1973. The proposal's failure elicited dismay and protests in Guatemala City. The Guatemalan National Committee for the Blind and Deaf, medical organizations, and other advocacy groups rallied demonstrations in support of efforts to provide vitamin A through fortification of sugar [43].

Finally, Guatemala's first sugar fortification law was enacted in July 1974. The decision was influenced by a decree for sugar fortification in neighboring Costa Rica, issued by their president, José Figueres Ferrer, in April 1974. Sugar fortification began in Guatemala in late 1975. The sugar fortification law required that the sugar producers absorb the costs of the fortification, although not all complied [43]. Twelve rural communities were the first to receive vitamin A-fortified sugar, with evaluations, done by blood tests, at six-month intervals throughout 1977 [44]. Vitamin A intake nearly tripled over that period, and the proportion of rural preschool children whose blood showed a vitamin A deficiency dropped in the course of a year from about 20 to 5 percent [42].

Encouraging results did not assure the Guatemalan sugar-fortification program's future, however. Some sugar producers continued to complain that government was interfering with private enterprise. The price of vitamin A itself rose, and the banks would not release foreign currency for the purchasing of vitamin A abroad (the materials required for fortification were unavailable in Guatemala) [45]. Within a few years the vitamin A program was effectively scrapped through negligence and as a result of resistance from the sugar industry. By the mid-1980s, vitamin A deficiency in Guatemala was back where is had been in the 1960s [43].

Two more decades passed before INCAP, the Ministry of Health, and UNICEF renewed efforts to reinstate the vitamin A fortification of sugar in Guatemala. This time the sugar producers participated in planning the program, which was restarted in 1988. As a result, fortified sugar reached 95% of Guatemalan households.

While Guatemala's sugar-fortification program found, lost, and finally recovered acceptance in its home country, the international health community was confronting the developing world's vitamin A deficiency problem head on. At the World

Food Conference, held in 1974 at the UN Food and Agriculture Organization (FAO) headquarters in Rome, the attendees resolved to urge a global reduction in vitamin A deficiency: 'Governments should. . . establish a world-wide control programme aimed at substantially reducing deficiencies of vitamin A (and other micronutrients). . .as quickly as possible' [46]. In the following year, the International Vitamin A Consultative Group (IVACG) was founded with support from the US Agency for International Development (USAID). The new group's purpose was to provide a forum for exchange of ideas and research findings on vitamin A, and to give technical guidance to policymakers and program managers [47].

Far from Latin America, other countries were also beginning to use high dose vitamin A supplementation to reduce vitamin A deficiency. In portions of India, a national program begun in 1971 provided all children between ages one and five with 200,000 IU of oral vitamin A every six months. The program was implemented in India's eastern and southern states where vitamin A deficiency was highly prevalent. Staff members of the Primary Health, Mother and Child Health, and Family Planning centers carried local responsibility, and auxiliary nurse-midwives gave vitamin doses during their house-to-house visits [48]. (The program ran into some initial difficulties: the bottles of vitamin A syrup were labeled 'India Family Planning Program', so some villagers, assuming the contents would sterilize their children, refused to participate.)

Bangladesh and Indonesia, too, began vitamin A supplementation programs in 1973. Indonesia in particular clearly had a massive vitamin A deficiency problem. Many questions were raised about the causes of xerophthalmia, its relationship to infectious diseases, and the impact of social, economic, and environmental factors on vitamin A deficiency. According to a development specialist with Helen Keller International, a large non-governmental organization (NGO) dedicated to the prevention of blindness, 'Some consider that the true public health weight of the problem in Indonesia is obscured because its victims often die before they can be reported as blind' [49]. The Indonesian government gave vitamin A capsules of 200,000 IU in twenty districts in Java to children between ages one and four [50].

In 1975, Indonesia's Ministry of Health established a steering committee to explore the possibility of a national program for the prevention of nutritional blindness. What obstacles would such an effort confront? With support from USAID, the Ministry of Health and Helen Keller International together launched the Nutritional Blindness Prevention Project in 1976 with headquarters in the Cicendo Eye Hospital in Bandung on the island of Java. Alfred Sommer, an ophthalmologist and epidemiologist at the Wilmer Eye Institute of the Johns Hopkins School of Medicine, was appointed Project Scientist. The main Indonesian collaborator in the field studies would be Ignatius Tarwotjo, director of the Academy of Nutrition of the Ministry of Health. Among the project's first aims were to measure the incidence of xerophthalmia and its prevalence among Indonesia's preschool children. To pursue these goals, the team studied four thousand six hundred children up to age six living in rural villages in the district of Purwakarta, which was known to have many children with xerophthalmia. From March 1977 to December 1978, the physicians examined the children every three months. This provided data on how common the disease was among these children and how many new cases appeared over time.

At roughly the same time, the Indonesian government, influenced by the Guatemala's ultimately successful experience, also began to consider vitamin A fortification of foods. As in Guatemala, the right food vehicle to be fortified had to be found. The choice, based on the usual Indonesian diet, was wheat flour, sugar, or monosodium glutamate (MSG). Of the three, only MSG was considered a suitable vehicle, as this food flavor-enhancer additive was inexpensive and widely sold throughout the Indonesian archipelago. (Government officials initially voiced concerns that using MSG as the vitamin A-fortification substance could give the appearance of official endorsement of a commercial product.) Scientific analysis, based on the assumption that providing vitamin A improves child survival, persuaded many participants in the effort that MSG was the optimal medium: even if only 10 percent effective, fortification of MSG with vitamin A would be a cost-effective way to save the lives of some twenty thousand children each year [51].

After completion of the fieldwork of the Nutritional Blindness Prevention Project in 1979, Sommer returned to the United States, where he continued to analyze the data collected in Indonesia. His results suggested that children who had night blindness, Bitot's spots, or both – symptoms of so-called 'mild' xerophthalmia – had a fourfold higher chance of dying than did children who were free of xerophthalmia. Working with three colleagues, Sommer showed that 'mild' vitamin A deficiency was associated with at least 16 percent of all deaths in children ages one to six. The team's conclusions: '. . .mild xerophthalmia justifies vigorous community-wide intervention, as much to reduce childhood mortality as to prevent blindness, and that night blindness and Bitot's spots are important as anthropometric indices in screening children to determine which of them need medical and nutritional attention' [52]. In other words, the term 'mild', attached to either xerophthalmia or vitamin A deficiency, was proving to be a mischaracterization. Children with 'mild' vitamin A deficiency were in fact at high risk of dying (textbox 8–3).

Textbox 8–3. H.A.P.C. Oomen: right too soon

Although 'Janus' Oomen, through the depth of his clinical experience, declared in 1958 that 'mild' xerophthalmia was responsible for a large proportion of child mortality in the community where he worked, he could not support his contention without the institutional backing and the extensive epidemiological data that Alfred Sommer and colleagues obtained two decades later. Oomen believed that the second year of life was the most dangerous year for the child in Indonesia; this notion was proven to be correct, as Sommer and colleagues demonstrated: the death rates were highest for children in the second year (fig. 8.2).

For the scientists working in nutrition and public health, observational data were suggestive of causal relationship: if a child had so-called 'mild' vitamin A deficiency, the child was at increased risk of dying as a result of the deficiency. To prove a causal relationship, however, the ultimate standard was a clinical trial that could show that giving vitamin A could prevent children from dying. The observations from Indonesia provided the stimulus for a major flurry of research on vitamin A that occurred over the following decade. The findings of Sommer and his colleagues' field studies in 1977–1978 in effect proved what Oomen had intuited from clinical practice two decades before.

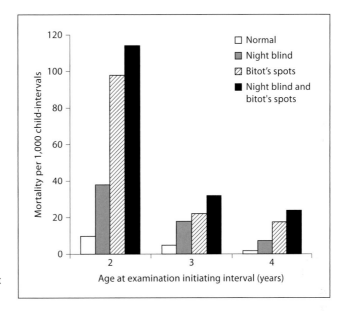

Fig. 8.2. Mortality among children who were free from respiratory disease and were 2–4 years of age at the examination initiating each of the six study intervals [52].

Experimental animal studies and clinical studies had firmly established by the 1960s that vitamin A played a role in normal immune function and in resistance to many infectious diseases. Practical experience in the field had borne out these findings. In response to the mounting evidence, WHO recommended the creation of a committee to undertake a systematic review of the scientific literature. Three academics led the effort. Nevin Scrimshaw of the Massachusetts Institute of Technology, Carl Taylor of Johns Hopkins University, and John Gordon of Harvard University, reviewed the large body of clinical and experimental evidence that had accumulated. The result, published in 1986, was an influential study that became a classic in nutrition, *Interactions of Nutrition and Infection* [53]. The work was instrumental in providing the biological underpinnings for the next decisive phase in the public health efforts to understand and control vitamin A deficiency in developing countries.

References

1 Santos, J., dos (1609) Ethiopia Oriental, e vária história de cousas, notáveis do Oriente. Évora, Manoel de Lyra; Jackson, J. G. (1811) An account of the Empire of Marocco, and the districts of Suse and Tafilelt: Compiled from miscellaneous observations made during a long residence in, and various journies through, these countries. 2nd edition. London, W. Bulmer; Bondt, J. de (1642) De medicina indorum, lib. IV. Amsterdam, Lugduni Batav; Hillary, W. (1759) Observations on the changes of the air and the concomitant epidemical disease, in the island of Barbados. To which is added a treatise on the putrid bilious fever, commonly called the yellow fever; and such other diseases as are indigenous or endemical, in the West India islands, or the torrid zone. London, C. Hitch and L. Hawes; Pisonis, G. (1648) Historia naturalis brasiliae. De medicina brasiliae. Libri quator: I. De aëre, aquis, & locis. II. De morbis endemiis. III. De venenatis & antidotes. IV. De facultatibus simplicium. Amsterdam, L. Batavorum; Ullersperger, B. [Account by Gama Lobo reported by Ullersperger and translated to German]. Brasilianische Augenentzündung (Ophthalmia braziliana). Klinische Monatsblätter für Augenheilkunde 1866;4:65–75. Also: Gazeta médica de Lisboa 1865; No. 16, 28 August 1865, 430, and No. 17, 13 September 1865, 466; de Gouvêa, H. (1882) Contribuição para o estudo da hemeralopia e a xerophthalmia por vicio de nutrição. Gazeta médica brazileira 1882;1:13–16, 67–72, 92–97, 139–145, 212–222; Entrecolles, F. X. (1781) Lettre du Père d'Entrecolles, Missionaire de la Compagnie de Jésus, au Père Duhalde, de la Même Compagnie. Peking, 8 October 1736. Jésuites. Lettres édifiantes et curieuses, écrites des missions étrangères. Nouvelle Édition, vol 22, Paris, J. G. Merigot, 183–245.

2 Ouwehand, C. D. (1900) Over 'rondar manok'. Geneeskundig tijdschrift voor Nederlandsch-Indië 40, 227–229.

3 Wille, W. A. (1922) Xerophthalmia, a deficiency disease (avitaminosis). In Far Eastern Association of Tropical Medicine. Transactions of the Fourth Congress held at Weltevreden, Batavia (Dutch East Indies), 1921. Weltevreden, Javasche Boekhandel en Drukkerij, 245–257.

4 de Haas, J. H. (1931) On keratomalacia in Java and Sumatra (in particular upon the Karo-Plateau) and in Holland. Mededeelingen dienst der volksgezondheid in Nederlandsch-Indië 20, 1–11.

5 Tijssen, J. (1927) Oorzaken van blindheid in Atjeh. Geneeskundig tijdschrift voor Nederlandsch-Indië 67, 99–101.

6 Tijssen, J. (1934) De invloed van xerophthalmie en andere veel voorkomende oogziekten op het aantal blinden in een land. Geneeskundig tijdschrift voor Nederlandsch-Indië 78, 5452–5458; Tijssen, J. (1939) De invloed van de xerophthalmie op het blinde-jongens-overschot in Nederlansch-Indië. Geneeskundig tijdschrift voor Nederlandsch-Indië 79, 79–83.

7 Hoat, O. D. (1936) Het voorkomen van xerophthlamie in eenige desa's van het regentschap Keboemen. Geneeskundig tijdschrift voor Nederlandsch-Indië 76, 1101–1110; Lian, S. B. (1937) Zijn er inderdaad meer mannen dan vrouwen blind tengvolge van xerophthalmie? Geneeskundig tijdschrift voor Nederlandsch-Indië 77, 3283–3286; Hadikoesoemo, G. A. (1938) Nogmaals over xerophthalmie. Geneeskundig tijdschrift voor Nederlandsch-Indië 78, 935–941; Maäs (1939) Xerophthalmie onder de Indonesische bevolking van het gewest Palembang. Geneeskundig tijdschrift voor Nederlandsch-Indië 79, 1512–1522.

8 van Veen, A. G., Postmus, S. (1947) Vitamin A deficiencies in the Netherlands East Indies. Journal of the American Dietetic Association 23, 669–673.

9 de Haas, J. H., Meulemans, O. (1938) Vitamine A en carotinoiden in bloed. II. Over het vitamine A- en carotinoiden-gehalte in het bloed van zwangere en zoogende inheemsche en Chineesche vrouwen te Batavia. Geneeskundig tijdschrift voor Nederlandsch-Indië 78, 847–855; de Haas, J. H., Meulemans, O. (1940) Vitamine A en caroteinoiden in bloed. II. Over het vitamine A- en carotinoidengehalte in het bloed van zwangere en zoogende inheemsche en Chineesche vrouwen te Batavia. Geneeskundig tijdschrift voor Nederlandsch-Indië 80, 928–950; Donath, W. F., Gorter, F. J. (1938) Vitamine A- en C- en carotinoidenbepalingen in het bloed van de dessabevolkirg van de gebieden rond Grissee, Segalaherang en Tjiandjoer. Geneeskundig tijdschrift voor Nederlandsch-Indië 78, 2235–2274; van Eekelen, M., de Haas, J. H. (1934) Over carotene en vitamine A in moedermelk, in het bizonder in colostrum. Geneeskundig tijdschrift voor Nederlandsch-Indië 74, 1201–1208; de Haas, J. H., Meulemans, O. (1937) Over het gehalte aan carotene en vitamine A van koemelk en koemelk-mengsels (een bijdrage tot de vitamine A-prophylaxe). Geneeskundig tijdschrift voor Nederlandsch-Indië 77, 279–288.

10 Lian, S. B. (1937) Onderzoekingen betreffende het vitamine A gehalte van het bloed van patiënten met oogsymptomen van avitaminose-A. Geneeskundig tijdschrift voor Nederlandsch-Indië 77, 1786–1794; de Haas, J. H., Meulemans, O. (1938) Vitamin A and carotinoids in blood. Deficiencies in children suffering from xerophthalmia. Lancet i, 1110–1111.

11 Meulemans, O., de Haas, J. H. (1936) Over het gehalte aan carotene en vitamine A van moedermelk in Batavia. Geneeskundig tijdschrift voor Nederlandsch-Indië 76, 1538–1570.

12 Straub, M. (1928) Kindersterfte ter oostkust van Sumatra. Koninklijke Vereeniging Koloniaal Instituut, Amsterdam. Mededeeling no. 24. Afdeeling Tropische Hygiëne, no. 15, 163–173; Sampoerno (1928) Steenvorming bij Avitaminose van den mensch. Geneeskundig tijdschrift voor Nederlandsch-Indië 68, 579–587; Wille, W. A. (1933) Nieuwe ondervindingen omtrent keratomalacie. Geneeskundig tijdschrift voor Nederlandsch-Indië 73, 279–285; Lian, S. B. (1933) Over xerophthalmie bij voldoende toevoer van vitamin A. Geneeskundig tijdschrift voor Nederlandsch-Indië 73, 105–109; Tijssen, J. (1936) De avitaminose van het oog en hare oorzaken. Geneeskundig tijdschrift voor Nederlandsch-Indië 76, 2891–2898.

13 Lian, S. B. (1929) Avitaminose A bij Inlandsche zuigelingen. Geneeskundig tijdschrift voor Nederlandsch-Indië 69, 1097–1103.

14 den Hartog, A. P. (2001) Acceptance of milk products in southeast Asia: the case of Indonesia as a traditional non-dairying region. In Cwiertka, K., Walraven, B. (eds) Asian food: the global and the local. Honolulu, University of Hawaii Press, 34–45.

15 Deeks, W. E. (1924) The use of sweetened condensed, evaporated and powdered milks for feeding infants in the tropics. American Journal of Tropical Medicine 4, 113–130.

16 Donath, W. F. (1938) Nogmaals afgeroomde gesuikerde melk. Geneeskundig tijdschrift voor Nederlandsch-Indië 78, 1258–1267.

17 de Haas, J. H., Meulemans, O. (1937) Over het gebruik van blikkenmelk in Ned.-Indië. Geneeskundig tijdschrift voor Nederlandsch-Indië 77, 1168–1186.

18 Donath, W. F. (1930) De voedingswaarde van blikkenmelk; onderzoek naar het gehalte van anti-xerophthalmie-, anti-beri-beri- en anti-scorbutvitamine. Geneeskundig tijdschrift voor Nederlandsch-Indië 70, 129–139.

19 de Kadt, G. S. (1937) Eenige bijzonderheden omtrent de voedingswaarde van gecondenseerde gesuikerde ondermelk naar aanleiding van het artikel van Dr. J. H. de Haas en Ir. O. Meulemans 'Over het gebruik van Blikkemelk in Ned.-Indië'. Geneeskundig tijdschrift voor Nederlandsch-Indië 77, 2818–2826.

20 de Haas, J. H. (1937) Qui s'excuse s'accuse. Geneeskundig tijdschrift voor Nederlandsch-Indië 77, 2827–2828; Baart de la Faille, J. M. (1937) Over het gebruik van blikkemelk in Ned.-Indië. Geneeskundig tijdschrift voor Nederlandsch-Indië 77, 3060–3061.

21 de Haas, J. H., Posthuma, J. H., Meulemans, O. (1940) Xerophthalmie bij kinderen in Batavia. Geneeskundig tijdschrift voor Nederlandsch-Indië 80, 928–950.

22 Williamson, A. D., Leong, P. C. (1979) Keratomalacia in Singapore and its relation to vitamin A in milk. Medical Journal of Malaya 4, 83–95.

23 Stanton, J. (1992) Obituary: Dr. Cicely Williams. The Independent, July 16, 1992.

24 PP/CDW/B.2/2. Williams, C. D. (1939) 'Milk and Murder' speech to Singapore Rotary Club.

25 American Public Health Association (1982) Policy statements adapted by the Governing Council of the American Public Health Association, November 4, 1981. Nestlé boycott. American Journal of Public Health 72, 205.

26 van Voorst, B., Andersen, K. (1981) The battle of the bottle: in Geneva it was the US against the world. Time June 1, 1981.

27 Sandstead, H. H. (2005) Origins of the Interdepartmental Committee on Nutrition for National Defense, and a brief note concerning its demise. Journal of Nutrition 135, 1257–1262.

28 Interdepartmental Committee on Nutrition for National Defense (1964). The Hashemite Kingdom of Jordan Nutrition Survey on Infants and Preschool Children in Jordan, November 1962–October 1963. National Institutes of Health, Bethesda; Interdepartmental Committee on Nutrition for National Defense (1959) Ethiopia Nutrition Survey. Bethesda, National Institutes of Health; Interdepartmental Committee on Nutrition for National Defense (1960) Republic of Vietnam Nutrition Survey, October–December 1959. Bethesda, National Institutes of Health; Interdepartmental Committee on Nutrition for National Defense (1962). The Kingdom of Thailand Nutrition Survey, October–December 1960. Bethesda, National Institutes of Health; Interdepartmental Committee on Nutrition for National Defense (1962). Republic of Lebanon Nutrition Survey, February–April 1961. Bethesda, National Institutes of Health; Interdepartmental Committee on Nutrition for National Defense (1966) Pakistan: Nutrition Survey of East Pakistan, March 1962–January 1964. Bethesda, National Institutes of Health.

29 Kinney, T. D., Follis, R. H. Jr. (eds) (1958) Nutritional Disease: Proceedings of a Conference on Beriberi, Endemic Goiter and Hypovitaminosis A, held at Princeton, N.J., June 1–5, 1958. Federation Proceedings 17 (Supp 2), i–viii, 1–162.

30 Oomen, H. A. P. C. (1953) Infant malnutrition in Indonesia. Bulletin of the World Health Organization 9, 371–384.

31 Oomen defined mild xerophthalmia as conjunctival xerosis and Bitot's spots in this paper.

32 Oomen, H. A. P. C. (1958) Clinical experience on hypovitaminosis A. In Kinney & Follis (1958), pp. 120–121.

33 Oomen, H. A. P. C., McLaren, D. S., Escapini, H. (1964) Epidemiology and public health aspects of hypovitaminosis A. A global survey on xerophthalmia. Tropical and Geographical Medicine 4, 271–315.

34 György, P., Burgess, A. (eds) (1965) Protecting the pre-school child: programmes in practice. Report on an international symposium held at the Rockefeller Villa Serbelloni, Lake Como, 3–8 August 1963. London, Tavistock.

35 National Academy of Sciences (1966) Pre-school child malnutrition: primary deterrent to human progress. An international conference on prevention of malnutrition in the pre-school child, Washington, D.C., December 7–11, 1964. National Academy of Sciences – National Research Council Publication 1282. Washington, D.C., National Academy of Sciences.

36 McLaren, D. S. (1966) The prevention of xerophthalmia. In National Academy of Sciences (1966), 96–101.

37 György, P. (1966) Programs for combating malnutrition in the pre-school child in Indonesia. In National Academy of Sciences (1966), 105–111.

38 Wolbach, S. B., Howe, P. R. (1925) Tissue changes following deprivation of fat soluble A vitamin. Journal of Experimental Medicine 42, 753–777; Richards, M. B. (1935) The role of vitamin A in nutrition. British Medical Journal i, 99–102; Tilden, E. B., Miller, E. G. Jr. (1930) The response of monkey (Macacus rhesus) to withdrawal of vitamin A from the diet. Journal of Nutrition 3, 121–140; Pillat, A. (1929) Does keratomalacia exist in adults? Archives of Ophthalmology 2, 256–287, 399–415; Sweet, L. K., K'ang, H. J. (1935) Clinical and anatomic study of avitaminosis A among the Chinese. American Journal of Diseases of Children 50, 699–734; Rit, P. (1937) Treatment of infantile diarrhoea with vitamin 'A'. Calcutta Medical Journal 32, 454–457.

39 Ramalingaswami, V. (1948) Nutritional diarrhoea due to vitamin A deficiency. Indian Journal of Medical Sciences 2, 665–674.

40 Pan American Health Organization (1970) Hypovitaminosis A in the Americas: Report of a PAHO Technical Group Meeting (Washington, D.C., 28–30 November 1968). Scientific Publication No. 198. Washington, D.C., Pan American Health Organization.

41 Sebrell, W. H. Jr. (1966) Population and food supply – implications in this hemisphere. In Proceedings Western Hemisphere Nutrition Congress organized by the Council on Foods and Nutrition, American Medical Association, November 8–11, 1965, Chicago, Illinois. Chicago, American Medical Association, 7–13.

42 Arroyave, G., Aguilar, J. R., Flores, M., Guzmán, M. A. (1979) Evaluation of sugar fortification with vitamin A at the national level. Scientific Publication 384. Washington, D.C., Pan American Health Organization and World Health Organization.

43 Mora, J. O., Dary, O., Chinchilla, D., Arroyave, G. (2000) Vitamin A sugar fortification in Central America: experience and lessons learned. Arlington, VA, MOST, The USAID Micronutrient Program.

44 Arroyave, G., Mejía, L. A., Aguilar, J. R. (1981) The effect of vitamin A fortification of sugar on serum vitamin A levels of preschool Guatemalan children: a longitudinal evaluation. American Journal of Clinical Nutrition 34, 41–49.

45 Solomons, N. W., Bulux, J. (1998) Vitamin A fortification survives a scare in Guatemala. Sight and Life Newsletter 2, 26–30.

46 Food and Agricultural Organization (1974) Assessment of the world food situation: present and future. Document C/Conf. 65/3, World Food Conference, Rome, FAO.

47 Reddy, V. (2002) History of the International Vitamin A Consultative Group 1975–2000. Journal of Nutrition 132, 2852S–2856S.

48 Kamel, W. W. (1973) A global survey of mass vitamin A programs. Washington, D.C. Office of Nutrition, Technical Assistance Bureau, Agency for International Development, US Department of State.

49 Fritz, C. (1980) Combating nutritional blindness in children: a case study in technical assistance in Indonesia. New York, Pergamon Press.

50 Vitamin A Deficiency Steering Committee (1980) Indonesia: nutritional blindness prevention project. Characterization of vitamin A deficiency and xerophthalmia and the design of effective intervention program. Final report. Jakarta, Helen Keller International.

51 Gershoff, S. N. (1982) Food fortification. In Scrimshaw, N. S., Wallerstein, M. B. (eds). Nutrition policy implementation: issues and experience. New York, Plenum Press, 61–71.

52 Sommer, A., Hussaini, G., Tarwotjo, I., Susanto, D. (1983) Increased mortality in children with mild vitamin A deficiency. Lancet ii, 585–588.

53 Scrimshaw, N. S., Taylor, C. E., Gordon, J. E. (1968) Interactions of nutrition and infection. Geneva, World Health Organization.

Chapter 9

Saving the Children: Rescue Missions against Strong Undertow

'If somebody has a political axe to grind, they will find a way to do it.'
Dr. Alfred Sommer, during the Albay Mother and Child Health Project (1987)

On July 30, 1986, in the lush, tropical Albay province in the of the Philippines' Bicol region, fieldworkers with the Albay Mother and Child Health Project began a house-to-house health survey. Loaded with bags, clipboards, and other equipment, the workers negotiated a precarious walk along the narrow, rough dirt paths separating rice paddies. In the distance, the nearly perfect cone of Mount Mayon emitted a thin, tranquil wisp that gave no hint of the cataclysms of which the volcano is capable. Its most recent eruption, in 1984, had forced the displacement of thousands of villagers.

Ideals for a New Era

With more than half its population of nearly four million people living in poverty, Bicol was one of the Philippines' poorest regions. Malnutrition was widespread there. The new survey was part of a two-week pilot study and practical training for the fieldworkers of the Albay Mother and Child Health Project, a research collaboration between the Johns Hopkins University, the Philippine Ministry of Health, and the Helen Keller International NGO. At all the houses they could reach, the fieldworkers interviewed mothers and gave all preschool children the standard vitamin A capsules that UNICEF has been distributing in developing countries since the 1970s to prevent or treat vitamin A deficiency [1]. For the previous fifteen years, UNICEF's vitamin A distribution efforts had been having impressive results in Bangladesh and India.

Village leaders and authorities from the Ministry of Health had been informed before the survey of the upcoming field activities. In early June, Alfredo R. A. Bengzon, the Philippines' Secretary of Health, had addressed the members of the Rural Health Units at the Daguisin-Dialogo Hall of the regional health office in Bicol's largest town, Legaspi City. The project launch came at a time of excited optimism in the Philippines. Corazon Aquino had been elected president on February 7, bringing to a close two decades of the corrupt dictatorship of Ferdinand Marcos and replacing it with 'people power.' (Indeed, Marcos had declared himself victor in the election, but his own military revolted and threw its support behind Aquino. On February 25, US military helicopters spirited Marcos and his family to Clark Air Force base and then to exile in Hawaii.)

Speaking to Rural Health Unit personnel, Bengzon explained to that, in the past, the main responsibilities of the Ministry of Health had been curative and preventive services, plus the implementation of public health programs. Under his care, the ministry's mission would broaden with a new emphasis on research. The work in Albay province would realize the goals he envisioned:

If we just confine ourselves to taking care of people and forget about the necessity of research. . . there is a certain sense of dissatisfaction – healthy dissatisfaction. . .. (I)n the Ministry, for as long as I'm there, research will be an integral hallmark of our tenure. . . as important as taking care of people. . .. (My) second message is about health and disease (which) know no. . . geographic or national boundaries. The fact that the project is being carried out in Albay is only an accident of resource availability and allocation. . .. (T)he benefits or the lessons to be learned from this study have international dimensions so that. . . what you people at the Rural Health Units will do. . . in Albay will affect some child out there in Ethiopia or Mozambique or Brazil. . .. (S)o you are really not simply Ministry of Health workers, you are international workers. . .. (T)he work you are going to do is probably far more important (for the) . . . lessons to be learned in the field of nutrition. . . even in the field of medical and human physiology. . .. (W)e are fortunate in health that we do not have to contend with nationalities (or race). . . since the focus in health is the human being. . .. We look at the things that go to his core: his health, his concern, his dignity and our responsibility to help him attain a measure of meaningfulness as he goes through life. . .. (T)he success and failure of any health care system eventually depends on mechanisms of the grass roots. . .. (P)eople who go on a day-to-day basis dealing with mothers, children. . . people who visit homes, people who make sure practices are carried out, people who make sure that follow-ups are made. . .. Never underestimate the value of your work. . .. Godspeed and good luck [2].

An initial task for the project was to assess the extent and severity of vitamin A deficiency in Albay province. Dr. Eva Santos, a Philippine ophthalmologist, examined hundreds of children and found that more than 3% had night blindness, Bitot's spots, or corneal scars characteristic of vitamin A deficiency. Her findings confirmed what many nutritionists had suspected: vitamin A deficiency was a serious problem among Bicol's preschool children. Furthermore, Santos and her colleagues believed, that beyond the obvious, '(T)here must be. . . a much larger number of cases in which the deficiency in vitamin A, without producing the eye disease, is the cause of a diminished resistance to infections, of general debility, and of malnutrition.' This confluence of factors had been noted in Denmark more than six decades before [3], and it was now the case in Bicol. Poverty was everywhere, child mortality rates were high, and vitamin A deficiency was rampant.

In the two-week pilot start of the Albay Mother and Child Health Project, the fieldwork done accomplished its goals: nearly five thousand preschool children in

Fig. 9.1. Field staff in the Albay Mother and Child Health Project, Philippines, in 1986. Photograph courtesy of Joanne Katz.

forty *barangays* (the Philippines' smallest administrative units; in effect, villages) received UNICEF's vitamin A capsules. The need for the project was proven – vitamin A deficiency in the region was indeed widespread – and workers to carry out the project were trained. What they would conduct would be a full-scale trial of vitamin A supplementation in the barangays of the Bicol region (fig. 9.1). The question the project sought to answer was: would the health and survival of preschool children improve as a result of regular administration every four months of 200,000 IU vitamin A capsules?

Under the leadership of Alfred Sommer, then professor of ophthalmology at Johns Hopkins University and later, dean of its School of Public Health, and Dr. Florentino Solon, director of the Nutrition Center of the Philippines, a team of physicians and other scientists would conduct the large, placebo-controlled clinical trial (textbox 9–1). The project would be the culmination of more then two years' planning involving much travel between the United States and the Philippines, negotiation of collaborative agreements, selection of a study site, and precise preparation of the research protocol. The protocol required independent review by ethical review boards at Johns Hopkins and in the Philippines to assure the protection of the people who would participate in the research.

Textbox 9–1. One right needs another to be proven right

A paper, published in the May 1986 issue of the *Lancet*, reported the results of a clinical trial, the 'Aceh Study,' carried out by in Indonesia under the leadership of Dr. Alfred Sommer of Johns Hopkins and his Indonesian nutritionist colleague, Ignatius Tarwotjo. As the Albay Mother and Child Health Project would attempt to do, the Indonesia study tested the efficacy of regular administration of 200,000 IU vitamin A capsules to improve the survival rate of preschool Indonesian children [4]. The trial had been conducted because recent field research in Indonesia had shown that children with night blindness and/or Bitot's spots had a higher risk of dying than did children without these signs or symptoms [5]. This, in turn, had raised the question of whether provision of vitamin A to children in a community setting would improve their survival.

The Indonesia study, conducted from 1982 to 1984, involved nearly twenty-six thousand children from four hundred and fifty villages in the Aceh area at the northern tip of Sumatra. The results showed that the mortality rate of Indonesian village preschoolers who received vitamin A was 4.9 deaths per 1,000 children per year. The children who did not receive the treatment experienced a mortality rate of 7.3 deaths per 1,000. These findings suggested that vitamin A supplementation could reduce the mortality of preschool children by about one-third. While the results from Indonesia were provocative, further corroboration of these findings was needed in other populations. Hence, the Albay Mother and Child Health Project of 1986.

The study would be conducted under the oversight of an independent data and safety monitoring committee, which had authority to stop the trial immediately if its members noticed anything that could adversely affect participants' safety [6]. Early termination could occur if interim results raised safety concerns, experimental treatment showed clear benefit, or both treatment and control groups reacted to the treatment in the same way. Another important element of the project was informed consent, which could be granted by a mother, father, or guardian. Potential participants had to be made aware of the study's purpose, its procedures, and its possible risks and benefits. Unlike most of the earlier human trials such as those reviewed in previous chapters, participation in the Albay Mother and Child Health Project had to be voluntary [7].

The Best Laid Plans. . .

The best guidance for formulating health policy and specific treatment recommendations comes from large, community-based clinical trials, and these must be conducted according to the highest scientific and ethical standards. Personnel at all levels, from top supervisors to fieldworkers to specialists in assessing the validity of data, must be either seasoned professionals or, for newcomers to this kind of work, rigorously trained.

Properly conceived and executed, large clinical trials are expensive to run, and appropriate sums must be allocated to data collection, treatment, and analysis of trial data. A reliable, genuinely useful, controlled clinical trial therefore calls for several elements:

- sufficient funding and organizational support,
- strong personal leadership,
- clear organization and well-planned logistics,
- carefully recruited and well-trained fieldworkers cooperate effectively under a common regimen of discipline,
- clear communications that promote full comprehension at all levels,
- sensitivity to the cultural standards, practices, fears, and history of the people participating as subjects, and
- acceptance of the study by the community.

Once data collection is completed, many questions must be asked in the process of scrutinizing the data and evaluating conclusions. Were the participants correctly identified each time they were visited? Were the placebo pills and treatment pills identical in appearance and taste so that participants did not know whether they were in the experimental or the control group? Were the study investigators and field workers kept unaware of the participants' treatment/control assignments – in other words, was it a strictly double-blind trial? Did the participants actually take their pills? Did participants have access to, and possibly take, vitamin A from sources outside the study? Was the dropout rate the same in both treatment groups, and was it high enough to undermine the statistical validity of the study? Was the data entry accurate, or did any fieldworkers falsify their data?

The Albay Mother and Child Health Project began with a high profile in the local area. 'The project was the biggest act in town', Kate Burns remembered. An American nurse with several years' experience in public health in Africa, Burns directed the field operations in Albay. She prominently advertised for positions in the project, screened and interviewed hundreds of applicants, and eventually hired nearly eighty local men and women to serve on the field teams and as office staff in the headquarters set up in Legaspi [8]. In the months before study's launch, fieldworkers visited the villages in Albay province to number and map the houses.

The two-week pilot study went well. It showed that the fieldworkers were prepared to carry out the main project, which would involve forty thousand preschool children in two hundred fifty barangays in Albay province. A local sheet metal worker manufactured nearly thirty thousand black-and-white metal plates 'like bicycle license plates', Burns recalled, 'stamped with the identification number for each house. . .'. The fieldworkers posted the plates on the houses of people who had agreed to participate.

But on Thursday morning, August 21, 1986, the residents of Bicol awakened to unwelcome news: Nelson Arao, a popular radio announcer, told his listeners that more than two dozen children in Albay had fallen ill with nausea, vomiting, and diarrhea after being given a vitamin A capsule [9]. The source of the news about widespread adverse reactions to vitamin A was Francis Burgos, a young physician who headed the Bicol chapter of the Medical Action Group, which had been founded in response to perceived human rights violations during the Marcos administration. Arao then broadcast a Medical Action Group statement from Bicol challenging that American researchers were using the children of Albay province 'as guinea pigs'.

The Philippine Ministry of Health, for its part, was accused of 'compromising the lives and health of our own children' with their apparent 'unshakable faith in foreign enterprise'. Hearing this, the Medical Action Group queried, 'Must we wait until there are enough dead/disabled/sick children and mothers before we attempt to protect them?' [10]. In response, officials at the Ministry of Health quickly announced a press conference the following day to explain the research project.

Local radio announcers, newspaper journalists, and other interested parties attended the Ministry of Health press conference, at which Eva Santos and other physicians with the ministry explained the purpose of the project. Someone harangued the gathering that the United States was dumping unwanted vitamin A capsules in the Philippines, since American children did not have xerophthalmia. Someone else charged that the Americans were giving away vitamin A capsules free now in order to stimulate future demand. One journalist questioned: Why was the project was being paid for by the U.S. government? – which ignited murmurings in the crowd about 'an international conspiracy'.

All six Legaspi broadcasters covered the Ministry of Health's press conference, but the radio coverage was soon drowned out by reports and accusations from Francis Burgos and the Bicol Medical Action Group. Earlier in the day, Burgos and his associates had brought nearly twenty children with 'toxic effects' of vitamin A to the regional health office (textbox 9–2). A local pediatrician, Elma Cabrera, of the Albay Provincial Hospital, examined the children and found them all to be malnourished and showing signs of fever, chronic diarrhea, and upper respiratory infection – all symptoms unrelated to vitamin A toxicity (textbox 9–2). In fact, questions put to the parents revealed that most of the children already had these problems before being given vitamin A capsules [11]. But Cabrera's medical evaluation was largely ignored, and radio stations continued for several days to denigrate the project.

Textbox 9–2. The side effects that can muddy the picture

Side effects following vitamin A supplementation (200,000 IU) are known to occur in about one to five percent of preschool children in developing countries. They consist of mild, transient headache and vomiting and diarrhea occurring within a twenty-four hours of the administration of a dose of vitamin A [12]. The local

Medical Action Group in Bicol, however, had misrepresented a host of illnesses as side effects of vitamin A. These included respiratory disease, conjunctivitis, skin rashes, and intestinal worms [6] many of which occurred many days after administration of vitamin A, or, in other cases, were already present before the children received any vitamin A [11]. The two children who died from alleged vitamin A toxicity were found, after pediatricians' further investigation, to have been ill before being given any vitamin A.

The benefits of vitamin A supplementation in reducing the risk of dying are considered greatly to outweigh the low risks of mild side effects. The mild side effects of vitamin A supplementation can be compared with the side effects of other measures taken to improve the health and survival of children. Diphtheria-pertussis-tetanus (DPT) and oral poliovirus (OPV) vaccines are used as routine childhood immunizations throughout the developing world. About one third or more of children who receive DPT immunization develop redness or pain at the injection site, fever, irritability, or vomiting, and one of 1,750 vaccine doses produces convulsions [13]. One out of 750,000 children who receives OPV vaccine develops paralytic polio. Although the side effects from vaccines can be serious, the benefits of childhood vaccination are considered to outweigh the risks of not being vaccinated against common childhood diseases.

A week later, the Ministry of Health held a scientific symposium in which the scientific background and protocol for the project were presented to local physicians and health workers. More than eighty physicians from the Albay Medical Society attended, and at the end of the symposium, the society formally endorsed the Albay Mother and Child Health Project.

All the while, Burgos and other members of Bicol's Medical Action Group picketed outside the Ministry of Health with signs reading 'STOP U.S. VITAMIN A.' They also took photographs of fieldworkers [1] – in some cases the same workers who had just been intimidated by Arao's radio broadcast. One activist reportedly threatened the fieldworkers and urged them to resign [1, 14]. In addition, the fieldworkers were branded 'anti-Philippine' and 'puppets of the Americans' [6, 15].

In one barangay, a group of fieldworkers was intercepted, interrogated, and ordered to quit the project by members of the New People's Army, the regular armed forces of the Communist Party of the Philippines [6]. According to one report, 'the group's alleged "saving factor" was a fieldworker's membership in the militant League of Filipino Students in his college days' [16]. The communists had questions: why were the Americans numbering houses and mapping Albay villages? *The Drug Monitor*, a newsletter published by the Health Action Information Network, reported that 'the Bicol groups have expressed fears that the project could have doubled as an anti-insurgency plot of the government in collusion with American intelligence agents' [17]. Frightened by threats of bodily harm, many fieldworkers feared for their safety and slept on the veranda of Kate Burns's rented house instead of returning home for the night [8].

Despite the formal support for the project from the Albay Medical Society, the Albay Hospital Association, the Albay Pediatric Society, the governor of Albay Province, and the Ministry of Health, the attacks on the project continued in the press and on the radio – largely fueled by Burgos and the Bicol Medical Action Group. Reports began to trickle back from the local Rural Health Units that mothers in the area were refusing vitamin A capsules for their children who had night blindness. Some mothers were questioning the safety of childhood immunizations against tetanus, diphtheria, and pertussis, asking whether the vaccines contained vitamin A [1]. On follow-up visits, fieldworkers began to find the metal house number plates they had posted torn down, tacked to conspicuous coconut palms, and marked with graffiti: R.I.P.

The number of Burgos children reported as having 'side effects from vitamin A' grew from two dozen to three dozen and finally, to sixty-five. On August 28, 1986, the news hit the capital when the *Manila Times* reported, 'Dr. Francis Burgos, head of the Albay Medical Action Group (MAG), said the research project called 'Albay Mother and Child Health Program' will virtually categorize children below six years as 'guinea pigs'' [18]. Similar stories alleging that the project was 'a conscious attempt to fool the people' appeared in local papers [19]. On Saturday, August 30, after receiving further threats from the Bicol Medical Action Group of picketing, court injunctions, and lawsuits, Kate Burns and two colleagues from Johns Hopkins loaded all the computers and printers into project vehicles and drove from Legaspi to Manila. The office was closed, and the fieldworkers were sent home. The activities of the project were put on an indefinite hold.

The furor did not end there, however.

Ten days later, the local Naga City newspaper in the Bicol region ran a story headlined 'Two Tots Died' – that is, died after receiving vitamin A [14]. In mid-September, the Ministry of Health and the national Medical Action Group formed an independent commission in Manila to investigate the research project and conduct an inquiry into what had now grown to eighty-five alleged cases of side effects and two child deaths from vitamin A [6]. The Bicol Medical Action Group would not budge and gave no facts or names regarding the eighty-five alleged vitamin A-caused illnesses and deaths.

At the end of September, a collaboration of the Ministry of Health and the national Medical Action Group commission, which included three vitamin A specialists, released a report. The conclusions supported the safety of the UNICEF vitamin A capsules used in the Albay project and endorsed the importance of the research being conducted. Bicol's Medical Action Group was not satisfied, of course. Not only did it reject the report; it also took the opportunity to re-invent itself.

In Bicol, the newly formed People's Committee to Scrap Vitamin A Project [6] found allies. The leftist New Patriotic Alliance, which had supported Corazon Aquino in the recent election, stepped into the fray. A political and not specifically health-oriented group, the New Patriotic Alliance saw the Albay Mother and Child Health Project as a manifestation of U.S. imperialism, since the project was funded by the U.S. Agency for International Development and the Ford Foundation [16]. Meanwhile, the agitation in Albay soon had unintended consequences for other programs. Catholic Relief Services, an international humanitarian agency, began to encounter resistance to its local community health efforts. The Ministry of Health was unable to move forward with implementation of the Tuberculosis Control Program in Albay, the last remaining province of the Philippines where the program was to be implemented [20].

The pendulum continued to swing. In the optimistic direction, further meetings were held during fall 1986 to negotiate a new beginning for the research project. A joint press release, issued by the Ministry of Health, the national Medical Action Group, and the Independent Investigation Committee, stated that the Albay Mother and Child Health Project was safe and important and should be implemented, with coordination at the local level with the Ministry of Health and local groups. The Bicol arm of the Medical Action Group abruptly pulled out of the planning and threatened further agitation.

In the pessimistic direction, the scientists involved concurred that continuation would be futile. Alfred Sommer and his Johns Hopkins colleagues knew the time had come to quit. 'We agonized over this', a rueful Sommer told a Philippine reporter. 'It shows you can't always cover all your bases. There's nothing we can think of that we can do in the future or that we would have done differently. If somebody has a political axe to grind, they will find a way to do it' [21]. (More besides scientific misperceptions and political agendas have obstructed the implementation of vitamin A supplementation efforts. Ideology too has impeded progress; see below on India's Dr. Colothur Gopalan.)

The planners of the Albay Mother and Child Health Project overestimated their preparedness in one of the essentials: community acceptance of the study. Their understanding of the Albay community had been incomplete, and the endorsement and cooperation of national and regional officials were perhaps misleading indexes of community support. They were blindsided by the actions of a disgruntled young physician and his anti-American sentiments; Francis Burgos skillfully used the media as his soap box [22]. Moreover, the influence of communism was gaining strength in the Philippines in the mid-1980s, particularly in Bicol province. Bicol has been particularly problematic in the recent history of the Philippines (textbox 9–3).

Textbox 9–3. Problematic Bicol

Though ideal from a scientific standpoint for a research project, Albay province offered a most inhospitable political environment. The province was ripe for the activities of the Communist Party and its rebel New People's Army of the Philippines. In the mid-1980s, the Communist Party controlled or influenced about 20% of the

Philippine population and wielded more than twenty-five thousand armed troops [23]. The group supported itself by levying 'taxes' on individuals and businesses, expecting peasant families in areas controlled by the New People's Army to give ten pesos and two cups of rice per month. A progressive tax was levied on businesses; nonpayment could result in labor problems or sabotage.

Other large projects also met resistance in Bicol. The New People's Army subsequently opposed a USD 50 million World Bank development project, the Community-Based Resource Management Project. The project aimed to reduce rural poverty and environmental degradation by increasing the capacity of local communities to implement and sustain natural resource management projects, improve environmental technology, expand environmental policies, and fortify the ability of local government to finance natural resource management projects [24]. The New People's Army stopped the implementation of the project in some Bicol municipalities by warning project staff members to cease their activities. They argued that similar World Bank projects in the past 'have only increased poverty incidence in their area of responsibility', and that the World Bank was 'known to be under control of the United States' [25].

Human rights reports show that the New People's Army has killed many civilians in the Bicol region. In the period 2000 to 2004, the highest number of killings and human rights violations was reported in Albay province [26]. Incidents such as the destruction of cell phone infrastructure were attributed to the inability of the biggest telecommunications companies to pay the annual 'revolutionary taxes' for each transmission site and for construction of new cell phone towers. The telecom companies are reportedly the largest sources for the 'revolutionary tax' of the New People's Army in Bicol. The group has also been known to burn buses and raze a local elementary school because extortion money was not paid. The construction of roads in the region has been hampered by the 'revolutionary taxes' needed for each kilometer of road building by the government. Local industries, such as marble cutting, were halted because of the 'taxes'. Parties whom the rebel group paid off have tolerated illegal logging and fishing. In 2004, fifty thousand people were displaced by the outbreak of armed conflict in Bicol [26].

More than a year after the Albay project had been shuttered, a former project staff member wrote to Kate Burns's colleague, Keith West:

I had been one of the enumerators of your Albay Mother and Child Health Project here in the Philippines. I wrote to tell you. . . that 'maybe' the trouble that led to the ruin of the project was indeed a blessing in disguise. Almost all of the project barangays. . . are very critical now, being totally infiltrated by the communists. Some of the barangays are now rebel camps. Our lives could have been endangered had the project continued. I am not happy that many kids are going blind around here. I daily see children with xerophthalmia, and just last month, a boy, totally blind died of a disease which can be. . . caused by lack of vitamin A. But the situation in our country is beyond control. Violence and criminality instantly soared sky-high. There's no use trying to help save children when anywhere unjust killings happen daily. We don't know what would happen to the children had they been helped by the project, and, maybe live for a few more years, then what? We simply don't know [27].

Getting It Right and on the International Agenda

In late 1987, Sommer and his Johns Hopkins colleagues established new ties for a controlled research trial to test the approach that had made the Indonesian Aceh Study successful (textbox 9–1). This time the points of contact were Dr. Ram Prasad Pokhrel, director of the Nepal Eye Hospital and the head of the National Society for the Prevention of Blindness, and Nepal's Ministry of Health. The ground in the flatlands below the Himalayas was well prepared for a large-scale vitamin A study designed along the lines of the aborted effort in the Philippines. Pokhrel and his colleague Larry Brilliant had conducted the Nepal Blindness Survey a few years earlier and found that vitamin A deficiency was indeed common in Nepal [28].

With Pokhrel, Sommer and West assembled a team to conduct a large, community-based trial of vitamin A supplementation in a rural, lowland region of southern Nepal known as the Terai. The area is part of the Ganges River floodplain that extends across northern India and runs through Bangladesh. Again, preparation for the study was painstaking, and nearly two years were devoted to the basic essentials: having the protocols reviewed and approved by ethical review committees in Nepal and the United States; hiring and training fieldworkers; establishing a data management center; and building communications with village leaders. The office equipment purchased for the Albay project was taken out of storage in the Philippines and transported to Nepal. More than fifty local workers were hired to enumerate the houses and follow the thousands of children who would be involved in the study. And as for local acceptance – one example is a man who owned a sugar mill. Upon learning that the study's aim was to improve child survival, he became keen to assist the project because his young son had recently died. The mill owner's support took the practical form of office space for the project and electricity generated at the mill.

The research project was officially launched at a ceremony held on May 27, 1989. Officials from the Ministry of Health, local dignitaries, and health personnel attended and were photographed flanking a large metal sign with a logo of the project and lettering in English and Nepali (fig. 9.2). A Hindu priest blessed the project before the participants, who were gathered in a large tent. But the formalities had hardly begun when the sky turned a strange yellow hue, signaling the advance of a huge windstorm from the west. The tent whipped around and then collapsed. Loudspeakers and folding chairs went flying. The participants ran for shelter through the rain torrent that followed. The large metal placard came crashing down, though nobody was injured. 'We thought the storm was a terrible sign', Sommer later reminisced, 'but the local people were happy, as it had been dry, and the storm indicated that the monsoon

Fig. 9.2. Some study team members of the vitamin A project in front of unfinished project offices, Sarlahi, Nepal, in 1988. From left to right, Sharada Ram Shrestha, B. D. Chataut, Keith P. West, Jr., Shankar Kedia, Alfred Sommer, and Nurnath Acharya. Photograph courtesy of Keith P. West, Jr.

rains were now coming' [29]. In this instance, a rainstorm boded well, though such was not always the case (textbox 9–4).

Textbox 9–4. An unsparing act of God

Sudan and Harvard University scientists had established a large vitamin A trial in the area of Khartoum. The study villages were located in five rural councils: Abu Dileig, Rifi Shamal, Rifi Genoub, and El Jaeli in Khartoum Province and Abu Haraz in El Gezira Province. The trial commenced in June 1988, but two months later, disaster struck the region in the form of unprecedented rains and flooding in Khartoum and surroundings areas [30]. A record 210 mm (8.25 inches) of rain fell within twenty-four hours, leading first to massive ground flooding and then to more flooding from the River Nile [31]. Upstream, dams could not restrain the waters of the Blue Nile swollen by rainfall in the Ethiopian highlands. More than six hundred of the region's villages disappeared completely, and some one-hundred thousand unbaked-clay houses were reduced to rubble [32]. One and a half million people were left without food or shelter [33]. The government declared a six-month month national State of Emergency [34]. Agencies and NGOs such as the Red Cross, Red Crescent, Doctors Without Borders, UNICEF, and Save the Children airlifted tents, medical supplies, water purification tablets, and food.

But the floods displaced the study participants living along the Nile in El Jaeli and Abu Haraz. Although the project planner had intended to provide placebo capsules instead of vitamin A to half the children, the mobile medical teams provided high-dose vitamin A capsules to all preschool children who had been displaced [35]. Rains continued through August. The UN reported that the Nile peaked at nearly seventeen meters above normal [36]. By the end of the month the relief operation had reached more than three hundred flights into the Khartoum airport. To add to the study's problems, the Sudan government, having declared a state of emergency, denied visas to the American investigators for eighteen months [37].

The results of the study, published in the *New England Journal of Medicine*, showed no significant impact of vitamin A supplementation on survival of children [38]. Nowhere was it mentioned in their paper that a large part of the study population was displaced in the middle of the trial or that the integrity of the placebo-controlled trial was compromised because children in the trial received vitamin A supplementation from emergency medical teams [39].

After an initial setback in the form of a disruption in fuel supplies for the project vehicles, the study got off to a strong start, with nearly thirty thousand children participating and the strong support of the community. From there, it proceeded much as the Albay Mother and Child Health Project should have, over a period of sixteen months, yielding ample and reliable results. In 1991, West, Pokhrel, Sommer, and colleagues published the Nepal study in the *Lancet* demonstrating that giving vitamin A supplements, 200,000 IU, every four months to preschool children reduced mortality by 30% [40].

The Nepal study corroborated the findings of the Aceh Study conducted in Indonesia (textbox 9–1), confirming the validity of a model that could be adapted for clinical trials in other regions where vitamin A deficiency was problematic. Even before the Nepal trial concluded, scientists elsewhere were hoping to repeat the success of the Aceh Study by conducting more trials in Indonesia, and well as in India, Ghana, and Sudan. Four additional studies from Asia showed that providing preschool children with vitamin A reduced child mortality [41]. One study conducted by scientists at India's internationally renowned Aravind Eye Hospital in Madurai showed that giving a community's preschool children weekly doses of vitamin A reduced child deaths by 54% [42].

Another study carried out in 1987 in Tanzania tested the effectiveness of treating children with measles with vitamin A and found a pronounced reduction in the severity of cases and occurrence of measles-caused deaths [43]. The Tanzania study was the first to corroborate the findings of the London pediatrician Joseph Ellison more than a half century before (chapter 7). Ellison's findings were confirmed once again by studies carried out in 1987 in South Africa [44]. With the results of many large trials the value of vitamin A to children exposed to measles finally won acceptance in the international health community. The World Health Organization and UNICEF issued a joint statement recommending high-dose vitamin A supplementation for all children diagnosed with measles who lived in communities where vitamin A deficiency was a recognized problem or where the proportion of children who died from measles was 1% or more [45].

In September 1990, at the World Summit for Children held at the United Nations in New York City, representatives from one hundred fifty-one countries endorsed the 'World Declaration for the Survival, Protection, and Development of Children'. The declaration recognized that an estimated forty thousand children were dying each day from malnutrition and disease. A Plan of Action that emerged from the summit set as its goal a worldwide one-third reduction of the death rate of children under age five. The plan also called for the 'virtual elimination of vitamin A deficiency and its consequences, including blindness' [46]. A policy conference, 'Ending Hidden Hunger', was held the next year in Montreal; representatives of more than fifty governments made a commitment to an intensive effort to fight micronutrient malnutrition.

Even with the accumulated results of many large trials, ambitious goals set, and commitments made, the need for scientific confirmation of the efficacy of vitamin A persisted. In 1992, the University of Toronto convened an international group of scientists to evaluate the studies that had been done and determine an overall assessment of the effect on morbidity and mortality in young children of vitamin A supplementation.

For a comprehensive answer to the question, 'Does vitamin A supplementation affect child mortality?' the group conducted a meta-analysis – a statistical technique that combines and integrates a collection of analytic results for the purpose of arriving at a single set of overarching conclusions. The meta-analysis, combining results from eight trials conducted in Ghana, India, Indonesia, Nepal, and Sudan, yielded the conclusion that vitamin A supplementation reduced child mortality by 23% [47]. A separate meta-analysis conducted by scientists at Harvard reached an even more persuasive conclusion: vitamin A supplementation reduced by 30% the overall mortality of a community's preschool children [48].

External Obstructions

Despite the international health organizations' proclamation that supplementary vitamin A went far in reducing child mortality, reinforced by incontrovertible statistical confirmation from meta-analyses, local forces in communities where vitamin A deficiency was taking a high toll could still obstruct the distribution of this sorely needed nutrient. In Guatemala, the vitamin A sugar fortification program that was enacted into law in 1974 (after much wrangling; chapter 8) proceeded smoothly from 1988 on until 1997, that is, when the price of sugar suddenly spiked a full 10%. Vitamin A deficiency had dropped significantly until then [49], but the jump in the sugar price nearly brought sugar fortification to a crashing halt. The Guatemalan government responded by dismantling the law mandating sugar fortification. On January 7, 1988, Vice President Luis Asturias announced an executive order that halted vitamin A supplementation in Guatemala.

In response, UNICEF and the Institute of Nutrition of Central America and Panama (INCAP) issued a warning that stopping the vitamin A fortification program would have adverse consequences for Guatemala, where 80% of the populace was living in poverty on vitamin A-poor diets [50]. The Guatemalan Committee for the Blind and the Deaf and private citizens as well, joined with UNICEF and INCAP to protest the government's decision. But the government's reaction to mounting criticism was to retrench: Asturias and his ministers held a press conference to restate the plan to rescind Guatemala's vitamin A fortification law. They offered a tentative consolation, however: many months would pass, they said, before children became blind. In the interim, some other measures to provide vitamin A might be put in place.

To the sugar industry, the notion of 'other measures' had an alarming ring: repeal of sugar fortification could do Guatemala's sugar producers no good. So a spokesman for the Guatemalan Association of Sugar Refiners – which in the past had vociferously opposed the use of sugar as the vehicle for delivering vitamin A – now reiterated the health institutions' warning that ceasing to fortify sugar would have adverse consequences for low-income Guatemalans. To bring his point home, the spokesman asserted that sugar offered the most cost-efficient means of assuring that the poor of Guatemala would have adequate vitamin A in their diets.

The issue brought human rights advocates, labor unions, and civic organizations together with the sugar magnates in opposition to dismantling the vitamin A-sugar fortification program. In addition, four congressmen and a private citizen filed a legal suit before the Constitutional Court claiming government's executive decision was unconstitutional: a law enacted at the will of the Legislature could not be overturned by executive order [50]. Government lawyers were given forty-eight hours to submit their briefs in opposition to the petition. After two tense days, the government backed down and resolved the crisis by means of announcing that the price of sugar would be returned to its 1997 pre-spike level, and mandatory fortification of sugar with vitamin A would continue.

Guatemala's vitamin A sugar fortification program survived subsequent scares. As in many other countries, sugar production in Guatemala is a highly protected domestic industry: the price of sugar on the international market, influenced mainly by competing supplies from such countries as Brazil and Cuba and demand from China and India, is usually lower than on the domestic market. In 2000, the international market price of sugar hit historic lows, creating a lucrative possibility for particularly enterprising Guatemalan sugar dealers who could purchase sugar cheaply on the international market and sell it within Guatemala for a much higher price – as long as the government agreed. Under Alfonso Portillo, the president who came into power in 2000, the Ministry of Economy gave the nod to this scheme and permitted the importation of cheap Cuban sugar. Predictably, protests greeted this decision, citing the likely loss of thousands of jobs in the domestic sugar industry. Furthermore, scientists with INCAP, personnel in public health institutions, and managers within the sugar industry itself reminded the government that all sugar in the country had to be fortified with vitamin A. Again, a nod from the government: the Ministry of Health reassured the public that the imported sugar would be fortified [51].

But the ministry had a surprise in store for the Guatemalan people. In September 2000, health minister Mario Bolaños denounced the country's sugar producers, declaring that they were not complying with national standards for vitamin A fortification of sugar [52]. The motive behind such allegations was to deflect criticism from the government's decision to allow sugar imports. But tests of the local product found that more than 80% had vitamin A concentrations that met the standards, and even those samples that fell short showed evidence of some fortification. (Under a collaborative external monitoring system established in 1995 and run by UNICEF, INCAP, and Guatemala's Ministry of Education, school children bring a sample of sugar from home to the public schools, where it is analyzed for its vitamin A concentration.) Omar Dary, a Guatemalan-born scientist who specializes in micronutrients and has worked throughout the developing world, remarked, 'Continued vigilance and monitoring are critical for any fortification program, because you cannot assume that business and government will do the right thing' [53].

Much Accomplished, More to Do

In light of the strong scientific evidence for vitamin A supplementation in reducing preschool child mortality, WHO and UNICEF increased their efforts starting in the late 1990s to strengthen national vitamin A supplementation programs. Through national programs in more than one hundred developing countries, children received vitamin A capsules every six months [54]. Tracking of various countries' programs by UNICEF has encouraged increased emphasis on child health and survival in developing countries. On September 8, 2000, one hundred eighty-nine UN member states adopted what was called the Millennium Declaration, with clear, time-bound goals for development. Millennium Development Goal Number Four has been of particular interest. It calls for a two-thirds reduction of under-age-five child mortality worldwide by 2015. An ambitious goal! To meet it will require a significant increase the coverage of vitamin A supplementation programs. In 2003 the programs altogether were reaching only 55% of the target population in developing countries [55].

Despite the progress made by public health programs initiated during the last half century, some ten million children are still dying each year in developing countries,

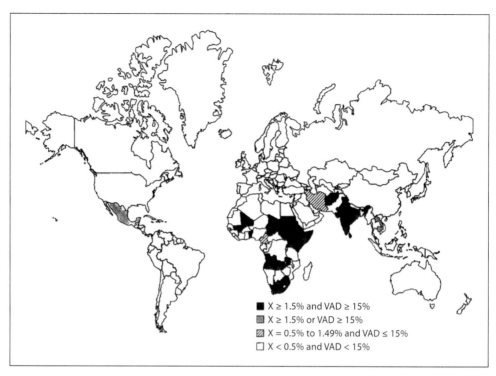

Fig. 9.3. Global distribution of vitamin A deficiency, defined by percent of children with low serum vitamin A levels (VAD) and with clinical signs/symptoms of vitamin A deficiency (X).

mostly of preventable causes [56]. More than 60% of those who die at age five and younger – the so-called 'under-five' deaths – could have been saved by implementation of a few known and proven interventions. Some one hundred twenty-five million preschool children around the globe are vitamin A-deficient – most of them living in southern Asia, Southeast Asia, and sub-Saharan Africa (fig. 9.3).

Vitamin A supplementation is a scientifically proven and practically tested means of improving child survival, and scientists affiliated with the World Bank and other organizations have found vitamin A supplementation to be one of the most cost-effective health interventions to reduce child mortality [57]. Economists who met in 2004 at the Copenhagen Consensus endorsed vitamin A supplementation as one of the best strategies to improve global well-being [58]. In the short 1998–2001 span alone, vitamin A supplementation prevented one million child deaths according to UNICEF [59]. Such countries as Indonesia and Vietnam, which have achieved high coverage with vitamin A supplementation, have noted a virtual disappearance of hospital admissions of children with night blindness, Bitot's spots, corneal ulceration and keratomalacia [60]. Vitamin A supplementation with 200,000 IU every four to six months has been shown to reduce the appearance of new cases of xerophthalmia among preschool children by 60–90% [61].

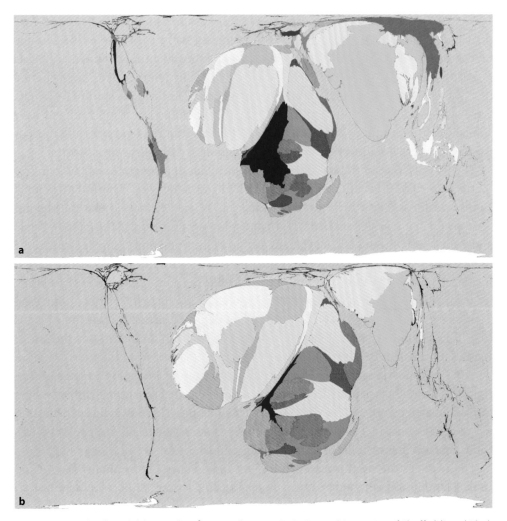

Fig. 9.4. a Under-five child mortality. ©Copyright 2006 SASI Group (University of Sheffield) and Mark Newman (University of Michigan). **b** Child deaths due to vitamin A deficiency. ©Copyright 2006 SASI Group (University of Sheffield) and Mark Newman (University of Michigan). Maps are cartograms (density equalizing maps) in which countries were resized according to under-five child mortality and child deaths from vitamin A deficiency, respectively.

But under-five child deaths persist at high rates even though this need not be the case, and nowhere is the problem worse than in India. With nearly two million under-five child deaths per year, India ranks at the top of all nations' under-five mortality [62]. India therefore figures prominently in any effort to reach the Millennium Development Goals. Shown on a map, India's share of global under-five deaths is rivaled only by that of sub-Saharan Africa (fig. 9.4). Compared with its immediate neighbors, India has made relatively little progress in mitigating child mortality since 1980 [63].

Bangladesh and Nepal, where under-five child death rates used to be even worse than India's, have long since surpassed India on the same measure [64]. The same holds true for comparisons with other neighbors on the Indian subcontinent. Analysts of health trends and the effectiveness of public health interventions have charted this discrepancy: while under-five child mortality fell in Bangladesh by 4.1%, India achieved only a 2.7 % reduction [65]. Given that nations of the subcontinent have similar histories, cultures, and institutions, the contrast in reduction of child mortality may be a result of differences in implementation of child survival strategies.

The low coverage of India's vitamin A supplementation program may account for that country's relative failure in reducing child mortality [64]. The India National Family Health Survey, which was conducted in 2005–2006, showed that the vitamin A supplementation program had 20% coverage among preschool children [64]. Nepal and Bangladesh, meanwhile, have managed to deliver vitamin A supplements to 85 to 90% of their preschool children. India, however, continues to experience a high incidence among children of night blindness, Bitot's spots, and blindness resulting from corneal ulceration and keratomalacia [66]. (According to the National Nutrition Monitoring Bureau, 0.7% of children in India under five years of age have Bitot's spots [67]; WHO defines vitamin A deficiency as a public health problem when the rate of Bitot's spots exceeds 0.5% [68].)

What has made the vitamin A deficiency situation in India so intractable? The probable answer is lack of national leadership in public health and nutrition, leading to inadequate coverage in vitamin A supplementation. There is a close correlation between high vitamin A deficiency, high under-five mortality, and deep poverty [69]. Sadly, this confluence of ills describes much of India.

The strongest index of poverty in India is landlessness: more than 40% of rural Indian families own no agricultural land [70]. With sixty million landless households, India has a greater landless population than any other country [71]. And of those households who do own some land, two hundred fifty million own less than 0.2 hectares (one-half acre). Many landless families survive mainly on cereals and only have one main meal a day [72]. But most rural land reforms in India have been quite weak and ineffective. Landless peasants have risen up in protest as part of a *Janadesh* (people's verdict) movement. In October 2007, some twenty-five thousand landless farmers – some wearing plastic bags as shoes – marched six hundred kilometers from the central Indian city of Gwalior to New Delhi to demand land [73].

Landlessness and poverty are not the only obstacles to improving the vitamin A status of Indian children. Ideology too stands in the way. One of India's most influential nutritionists, Dr. Colothur Gopalan, has long been instrumentally opposed to vitamin A supplementation and vitamin A fortification of foods. As director of the Nutrition Foundation of India, Gopalan commands attention when he urges that 'public-minded citizens must ensure the scrapping of the massive-dose vitamin A prophylaxis approach. This will not only avoid the considerable unnecessary expenditure which the government is incurring on the programme but more importantly, will save our children from undesirable side effects'. He argues that child mortality in developing countries should be reduced through the elimination of poverty (this argument has been tested before; textbox 9–5). Instead of vitamin A supplementation or fortification, Gopalan recommends that children eat more fruit and vegetables [74].

Umesh Kapil, a pediatrician who practices in Delhi, is one of Gopalan's followers. A Calcutta newspaper has quoted him as using arguments quite like Gopalan's against vitamin A supplementation. Kapil, too, would eradicate the problem only at the roots and suspects that opportunists are at work behind vitamin supplementation. '[W]e must look to our farmers, not to pharmaceutical companies, to protect the health of our children. The main solution to vitamin A deficiency should not be drug-based, but food-based. . .' and '. . .the issue of vitamin A has commercial overtones' [75].

The grim situation of India's landless peasants, most of whom subsist on less than the equivalent of one US dollar a day, makes advisories such as 'eliminate poverty' and 'eat more fruits and vegetables' seem either curiously naïve or cynical. In the context of India's poverty, landlessness, and child mortality, they seem little more than pathetic emulations of Marie Antoinette's infamous (and unproven), 'Let them eat cake'.

Textbox 9–5. The grand detour

Recent history had some hard lessons to teach the benevolent experimenters, agencies, and nongovernmental organizations which, in the 1980s, would introduce supplementation and food fortification to eradicate vitamin A deficiency and the dire health effects it caused. The experience of Spain offered a case in point. In some areas of Spain, the micronutrient iodine was severely lacking and causing an array of disorders very different from those caused by vitamin A deficiency but comparably damaging.

Dietary iodine is essential for normal function of thyroid hormones, and problems that can result from a lack of it include severe mental retardation ('cretinism'), impaired physical development (particularly of the brain), increased infant mortality, hypothyroidism, and goiter (enlargement of the thyroid gland) [76]. Iodine deficiency is a function of local presence of iodine in water, plants, and food. Mountainous regions and large river deltas are particularly prone to provide inadequate natural iodine, so populations in such terrain are especially vulnerable to iodine-deficiency disorders. Since the early part of the twentieth century, salt fortified with iodine (widely familiar as iodized salt) has been the primary strategy for preventing iodine deficiency. Iodized salt has been adopted almost worldwide.

During the mid-1900s, one of Spain's most influential physicians was Gregorio Marañon (1887–1960). As head of the Institute of Endocrinological Research and chair of Endocrinology at the University of Madrid, Marañon was well acquainted with the problems of goiter and cretinism which were widespread across many parts of Spain. In 1921, King Alfonso XIII created a Goiter Commission and appointed Marañon the head it, and in the following year, Marañon accompanied the king on a visit to Las Hurdes, a mountainous region in west central Spain with an extremely high prevalence of goiter and cretinism. Marañon urged the king to promote economic development in the area. He applauded the road construction, educational programs, and agricultural projects that soon began in the area, but he did not urge the iodization of salt [77]. The human suffering in Las Hurdes also caught the attention of filmmaker Luis Buñuel in his 1933 documentary, *Tierra sin Pan (Land without Bread)*, which was immediately banned by the Franco administration [78].

Marañon was convinced that economic development and improved diet would eradicate goiter and cretinism in Spain, including in places like Las Hurdes. Iodized salt was seen as too specific; the general problems underlying iodine deficiency would persist. What was needed to remedy the problem was the eradication of poverty itself, and with it, malnutrition. Marañon's was a philosophy of All or Nothing. Economic development was the only means to reduce poverty, disease, inbreeding, poor diet, and iodine deficiency [79]. He remained opposed to the iodization of salt until his death in 1960.

Heinrich Hunziger, a Swiss endocrinologist and proponent of iodized salt in Switzerland, referred to Marañon's high idealism as 'the Grand Detour' [80]. Hunziger showed that economic development alone did not lead to the eradication of iodine deficiency. Marañon was the most respected and influential endocrinologist in Spain, and, as noted by the historian Renate Fernandez, '(H)is prestige in the Spanish system of seniority and hierarchy may have insulated him from challenges by Spanish colleagues' [77].

In the years after Marañon's death, the Franco regime's restriction of medical information contributed to inaction with regard to iodine deficiency. The inertia lasted long past the Generalissimo's death in 1975. Iodized salt finally became available in Spain in 1981, but its use was not widespread and not compulsory [81]. Finally, in 1985, the European Thyroid Association published a study of endemic goiter in the countries of Europe, with results showing a shocking prevalence – a full 86% – of goiter in schoolchildren in Las Hurdes. This was the highest goiter rate in all Europe [82]. Marañon's 'Grand Detour' of economic development failed miserably: more than sixty years later, most of the children of Las Hurdes were still suffering from iodine deficiency. The resulting sickness and death that could have been prevented with the simple intervention of iodized salt in Spain is not known. The toll in terms of diminished capacity and human life was probably staggering. As for Marañon's lofty aspirations for his country, their echo still reverberates today in India and has the same harmful consequences.

More Vegetables and Fruit: Nice Idea, but. . .

Although increasing the consumption of vitamin A-rich foods may seem to be a reasonable solution, in reality, it is much more difficult for preschool children in poor families – even those families that hold some land – to meet the requirements for vitamin A through diet alone. Animal source foods that are rich in vitamin A, such as liver, eggs, cheese, and butter, are often beyond the reach of poor families. In the India National Family Health Survey, mothers were asked whether their children had consumed any vitamin A-rich foods (liver, fish, egg, dark green leafy vegetables, pumpkin, carrots, yellow or orange sweet potatoes, ripe mango, papaya, cantaloupe, and jackfruit) within the last twenty-four hours. (Twenty-four hour dietary recall is accepted as a valid dietary assessment method and is commonly used in nutrition surveys.) The results showed that more than 40% of children ages twelve months to thirty-five months did not receive any vitamin A-rich foods during the day preceding the interview [83]. Less than 8% of the children had received an egg. The national survey demonstrates that young children have a low consumption of vitamin A-rich fruit and vegetables and explains why vitamin A deficiency remains a deeply rooted public health problem in India.

Another critical factor that makes it difficult for preschool children to meet their dietary requirements for vitamin A through fruit and vegetables alone is that the bioavailability of vitamin A from fruit and vegetables is not high. (Bioavailability refers to the proportion of a nutrient contained in food that is actually absorbed by the body. For example, spinach leaves contains a certain concentration of vitamin A, but when spinach is eaten, not all its vitamin A is absorbed in the digestion process. In fact, the proportion absorbed is much lower.) In the early 1990s, studies conducted by Saskia de Pee, Clive West, and Muhilal in Indonesia showed that the bioavailability of vitamin A from fruit and vegetables is lower than was once believed [84]. The Institute of Medicine subsequently revised its guidelines regarding the bioavailability of vitamin A in fruit and vegetables [85]. The low bioavailability of vitamin A from vegetables has been corroborated by rigorous dietary studies in humans [86]. A young child between ages one year and three would need to eat eight servings of dark green leafy vegetables per day in order to meet the Recommended Dietary Allowance for vitamin A [87]. The problem of the low bioavailability of vitamin A in plant foods has brought the sobering reality of 'the virtual impossibility for most poor, young children to meet their vitamin A requirements through vegetable and fruit intake alone' [88]. The low bioavailability of vitamin A from plant foods explains, in part, the presence of vitamin A deficiency among children living amid ample supplies of dark green leafy vegetables and other plant sources of vitamin A.

In other words, the fight goes on. There have been many successes, but not enough. UNICEF continues to distribute more than five hundred million vitamin A capsules each year in developing countries and has saved the lives of millions of children since beginning its vitamin A program. But the challenges to total eradication of vitamin

A deficiency persist, largely as manmade ones. Donor fatigue has become a factor, and political obstacles still undermine efforts to improve child survival. So millions of children still suffer the effects of deficiency in the form of blindness, severe infections, and, for some, death. Efforts to lift completely the shadow of death cast by vitamin A deficiency – by now proven to be feasible and economical – in some places still entail struggle against intractable opposition. The victims who continue to suffer under the shadow are mainly the children.

References

1 JK. Ministry of Health, Republic of the Philippines. Summary report concerning adverse publicity for the Albay Mother and Child Health Project. August 27, 1986.

2 JK. Bengzon, A. Transcript of the address of Minister Bengzon to the members of the Rural Health Units at the Daguisin-Dialogo Hall of the Regional Health Office No. 5, Legaspi City. June 8, 1986.

3 Widmark (1924).

4 Sommer, A., Tarwotjo, I., Djunaedi, E., West, K. P. Jr., Loeden, A. A., Tilden, R., Mele, L., and the Aceh Study Group (1986) Impact of vitamin A supplementation on childhood mortality. A randomized controlled community trial. Lancet i, 1169–1173.

5 Sommer, A., Hussaini, G., Tarwotjo, I., Susanto, D. (1983) Increased mortality in children with mild vitamin A deficiency. Lancet ii, 585–588.

6 JK. Tielsch, J. Letter to M. Forman, January 9, 1987.

7 JK. Sommer, A. Letter to P. Johnson, September 2, 1986.

8 Burns, K. Personal communication, August 21, 2009.

9 JK. Burns, K. Letter to A. Sommer, September 15, 1986.

10 JK. Position paper: Our children are human beings too. Medical Action Group (MAG), Free Legal Assistance Group (FLAG), Albay's People Organization (APO), BICOLANDIA. August 21, 1986.

11 JK. Position paper: Our children's health. September 5, 1986.

12 World Health Organization/United Nations International Emergency Children's Fund/International Vitamin A Consultative Group Task Force (1988) Vitamin A supplements: a guide to their use in the treatment and prevention of vitamin A deficiency and xerophthalmia. Geneva, World Health Organization, p. 10; Sommer, A., West, K. P. Jr. (1996) Vitamin A deficiency: health, survival, and vision. New York, Oxford University Press, p. 394.

13 Cody, R. L., Baraff, L. J., Cherry, J. D., Marcy, S. M., Manclark, C. R. (1981) Nature and rates of adverse reactions associated with DTP and DT immunizations in infants and children. Pediatrics 68, 650–660.

14 JK. Burns, A. Letter to A. Sommer, September 15, 1986.

15 JK. Position paper: A rebuttal to the position paper against vitamin A. Albay Mother and Health Project, August 1986.

16 Cuezon, M. (1986) The continuing Albay vitamin A controversy. The Drug Monitor, No. 8, 1–8.

17 Cuezon, M. (1986) Albay vitamin A research project questioned. The Drug Monitor, No. 7, 1–6.

18 'Stop Albay research project, gov't urged.' Manila Times, August 28, 1986.

19 'Albay project stirs row.' Manila Bulletin, August 30, 1986; Icamina, P. Nutrition project raises issues of medical ethics. Manila Chronicle, November 24, 1986.

20 JK. Tielsch, J. Letter to A. Sommer and colleagues, November 25, 1986.

21 Icamina (1987).

22 Francis Burgos – the central figure behind the anti-American activities in Bicol – subsequently moved to the US. He opened a private practice in internal medicine in Phoenix, Arizona.

23 Kessler, R. J. (1989) Rebellion and repression in the Philippines. New Haven, Yale University Press, p. 28.

24 World Bank (2008) Implementation completion and results report (IBRD-42990) on a loan in the amount of US $50 million equivalent to the Republic of the Philippines for the Community-Based Resources Management Project. April 15, 2008. Report No. ICR0000723. Washington, D.C., World Bank.

25 Calara, P. M. (2002) $50-M WB project refuses to take off in Bicol. Bulattat vol. 2, no. 13, May 5–11, 2002.

26 Lobrigo, F. J. E., Imperial, S., Rafer, N. (2006) Armed conflict in Bicol: the price does not come cheap. Philippine Institute for Development Studies Policy Notes, No. 2006–4 (March 2006).

27 JK. Letter from former Filipino staff member to K. West, October 19, 1987.

28 Brilliant, L. B., Pokhrel, R. P., Grasset, N. C., Lepkowski, J. M., Kolstad, A., Hawks, W., Pararajasegaram, R., Brilliant, G. E., Gilbert, S., Shrestha, S. R., et al. Epidemiology of blindness in Nepal. Bulletin of the World Health Organization 1985;63:375–386.

29 Sommer, A. Personal communication, September 24, 2009.

30 United Nations Department of Humanitarian Affairs (1988) Sudan – flood. UNDRO Situation Report No. 1. August 6, 1988; Walsh, R. P. D., Davies, H. R. J., Musa, S. B. (1994) Flood frequency and impacts at Khartoum since the early nineteenth century. Geographical Journal 160, 266–279.

31 Sutcliffe, J. V., Dugdale, G., Milford, J. R. (1989) The Sudan floods of 1988. Hydrological Sciences 34, 355–364.

32 United Nations Department of Humanitarian Affairs (1988) Sudan – flood. UNDRO Situation Report No. 4. August 9, 1988.

33 United Nations General Assembly (1988) Emergency assistance to the Sudan. 18 October 1988, A/RES/43/08.

34 Hulme, M., Trilsbach, A. (1989) The August 1988 storm over Khartoum: its climatology and impact. Weather 44, 82–90.

35 United Nations Department of Humanitarian Affairs (1988) Sudan – flood. UNDRO Situation Report No. 9. August 22, 1988; Centers for Disease Control (1989) International notes health assessment of the population affected by flood conditions – Khartoum, Sudan. MMWR Weekly Report 37, 785–788.

36 United Nations Department of Humanitarian Affairs (1988) Sudan – flood. UNDRO Situation Report No. 12. September 2, 1988.

37 Sommer, A. (2008) Vitamin A deficiency and clinical disease: an historical overview. Journal of Nutrition 138, 1835–1839.

38 Herrera, M. G., Nestel, P., el Amin, A., Fawzi, W. W., Mohamed, K. A., Weld, L. (1992) Vitamin A supplementation and child survival. New England Journal of Medicine 340, 267–271.

39 The senior investigators did not respond to multiple e-mails and voice messages from the author requesting an opportunity to discuss details of their study in the Sudan.

40 West, K. P. Jr., Pokhrel, R. P., Katz, J., LeClerq, S. C., Khatry, S. K., Shrestha, S. R., Pradhan, E. K., Tielsch, J. M., Pandey, M. R., Sommer, A. (1991) Efficacy of vitamin A in reducing preschool child mortality in Nepal. Lancet 338, 67–71.

41 Muhilal, Permeisih, D., Idjradinata, Y. R., Muherdiyantiningsih, Karyadi, D. (1988) Vitamin A-fortified monosodium glutamate and health, growth, and survival of children: a controlled field trial. American Journal of Clinical Nutrition 48, 1271–1276; Rahmathullah, L., Underwood, B. A., Thulasiraj, R. D., Milton, R. C., Ramaswamy, K., Rahmathullah, R., Babu, G. (1990) Reduced mortality among children in southern India receiving a small weekly dose of vitamin A. New England Journal of Medicine 323, 929–935; Vijayaraghavan, K., Radhaiah, G., Surya Prakasam, B., Rameshwar Sarma, K. V., Reddy, V. (1990) Effect of massive dose vitamin A on morbidity and mortality in Indian children. Lancet 336, 1342–1345; Daulaire, N. M. P., Starbuck, E. S., Houston, R. M., Church, M. S., Stukel, T. A., Pandey, M. R. (1992) Childhood mortality after a high dose of vitamin A in a high risk population. BMJ 304, 207–210.

42 Rahmathullah (1990).

43 Barclay, A. J. G., Foster, A., Sommer, A. (1987) Vitamin A supplementation and mortality related to measles: a randomised clinical trial. British Medical Journal 294, 294–296.

44 Hussey, G. D., Klein, M. (1990) A randomized, controlled trial of vitamin A in children with severe measles. New England Journal of Medicine 323, 160–164.

45 Anon (1987) Vitamin A for measles. Lancet i, 1067–1068.

46 Horwitz, A. (1993) Global & political climate in which consequences of vitamin A deficiency & its prevention are being recognized. In Bellagio meeting on vitamin A deficiency & childhood mortality: Proceedings of – 'Public health significance of vitamin A deficiency and its control' Bellagio Study and Conference Center of the Rockefeller Foundation, February 3–7, 1992. New York, Helen Keller International, 7–12.

47 Beaton, G. H., Martorell, R., L'Abbé, K. A., Edmonston, B., McCabe, G., Ross, A. C., Harvey, B. (1992) Effectiveness of vitamin A supplementation in the control of young child morbidity and mortality in developing countries. Toronto, University of Toronto.

48 Fawzi, W. W. Chalmers, T. C., Herrera, M. G., Mosteller, F. (1993) Vitamin A supplementation and child mortality. A meta-analysis. JAMA 269, 898–903.

49　Dary, O., Martínez, C., Guamuch, M. (2005) Sugar fortification with vitamin A in Guatemala: the program's successes and pitfalls. In Freire, W. B. (ed.) Nutrition and an active life: from knowledge to action. Scientific and Technical Publication No. 612. Washington, D.C., Pan American Health Organization, 43–59.

50　Solomons, N. W., Bulux, J. (1998) Vitamin A fortification survives a scare in Guatemala. Sight and Life Newsletter 2/1998, 28.

51　González Pérez, L. E. (2000) Advierten riesgos de azúcar sin vitamina. Pulso Económico August 10, 2000, 41.

52　González Pérez, L. E. (2000) Denuncian hallazgo de azúcar mal vitaminada. Pulso Económico September 1, 2000, 40.

53　Omar Dary (October 19, 2009) Personal communication.

54　United Nations Children's Fund (2007) Vitamin A supplementation: a decade of progress. New York, UNICEF.

55　Jones, G., Steketee, R. W., Black, R. E., Bhutta, Z. A., Morris, S. S., and the Bellagio Child Survival Study Group (2003) How many child deaths can we prevent this year? Lancet 362, 65–71.

56　Black, R. E., Morris, S. S., Bryce, J. (2003) Where and why are 10 million children dying every year? Lancet 361, 2226–2234; United Nations Children's Fund. (2008) State of the World's Children 2009. New York, UNICEF.

57　World Bank (1993) World development report 1993: investing in health. New York, World Bank and Oxford University Press; Bryce, J., el Arifeen, S., Pariyo, G., Lanata, C. F., Gwatkin, D., Habicht, J. P., Multi-Country Evaluation of IMCI Study Group. (2003) Reducing child mortality: can public health deliver? Lancet 362, 159–164; Edejer, T. T. T., Aikins, M., Black, R., Wolfson, L., Hutubessy, R., Evans, D. B. (2005) Cost effectiveness analysis of strategies for child health in developing countries. BMJ 331, 1177.

58　Lomborg B (ed.) (2004) Global crises, global solutions. Cambridge, Cambridge University Press.

59　United Nations Children's Fund. (2001) Ending vitamin A deficiency: a challenge to the world. New York, UNICEF, 2001.

60　Semba (2007a), 50; Semba, R. D., Muhilal, Soesatio, B., Natadisastra, G. (1995) Decline of admissions for xerophthalmia to Cicendo Eye Hospital, Bandung, Indonesia, 1981–1992. International Ophthalmology 19, 39–42; Muhilal, Tarwotjo, I., Kodyat, B., Herman, S., Permaesih, D., Karyadi, D., Wilbur, S., Tielsch, J. M. (1994) Changing prevalence of xerophthalmia in Indonesia, 1977–1992. European Journal of Clinical Nutrition 48, 708–714; Bloem, M. W., Gorstein, J. (1995) Vietnam: xerophthalmia free. 1994 National Vitamin A Deficiency and Protein-Energy Malnutrition Prevalence Survey. Consultancy report, March 5–17, 1995. Jakarta, Helen Keller International.

61　Sinha, D. P., Bang, F. B. (1976) The effect of massive doses of vitamin A on the signs of vitamin A deficiency in preschool children. American Journal of Clinical Nutrition 29, 110–115; Djunaedi, E., Sommer, A., Pandji, A., Kusdiono, Taylor, H. R., the Aceh Study Group (1988) Impact of vitamin A supplementation on xerophthalmia. A randomized controlled community trial. Archives of Ophthalmology 106, 218–222; Sommer, A. (1995) Nutritional blindness: xerophthalmia and keratomalacia. New York, Oxford University Press; Katz, J., West, K. P. Jr., Khatry, S. K., Thapa, M. D., LeClerq, S. C., Pradhan, E. K., Pokhrel, R. P., Sommer, A. (1995) Impact of vitamin A supplementation on prevalence and incidence of xerophthalmia in Nepal. Investigative Ophthalmology and Visual Science 36, 2577–2583.

62　UNICEF (2008); Murray, J. L. M., Laakso, T., Shibuya, K., Hill, K., Lopez, A. D. (2007) Can we achieve Millennium Development Goal 4? New analysis of country trends and forecast of under-5 mortality to 2015. Lancet 370, 1040–1054.

63　Claeson, M., Bos, E. R., Mawji, T., Pathamanathan, I. (2000) Reducing child mortality in India in the new millennium. Bulletin of the World Health Organization 78, 1192–1199.

64　Semba, R. D., de Pee, S., Sun, K., Bloem, M. W., Raju, V. K. (2010) The role of expanded coverage of the national vitamin A program in preventing morbidity and mortality among preschool children in India. Journal of Nutrition 140, 208S-212S.

65　Murray (2007).

66　Gopaldas, T., Gujral, S., Abbi, R. (1993) Prevalence of xerophthalmia and efficacy of vitamin A prophylaxis in preventing xerophthalmia co-existing with malnutrition in rural Indian children. Journal of Tropical Pediatrics 39, 205–208; Khandait, D. W., Vasudeo, N. D., Zodpey, S. P., Ambadekar, N. N., Koram, M. R. (1999) Vitamin A intake and xerophthalmia among Indian children. Public Health 113, 69–72; Shaw, C., Islam, M. N., Chakroborty, M., Biswas, M. C., Ghosh, T., Biswas, G. (2005) Xerophthalmia: a study among malnourished children of West Mednipur District. Journal of the Indian Medical Association 103, 180, 182–183; Pal, R., Sagar, V. (2007) Correlates of vitamin A deficiency among Indian rural preschool-age children. European Journal of Ophthalmology 17, 1007–1009;Varughese, S. (2007) Vitamin A deficiency in children under 6 months. Tropical Doctor

37, 59–60; Arlappa, N., Laxmaiah, A., Balakrishna, N., Harikumar, R., Brahmam, G. N. (2008) Clinical and sub-clinical vitamin A deficiency among rural pre-school children of Maharashtra, India. Annals of Human Biology 35, 606–614; Karande, S., Jagtap, S. (2008) Ocular sequelae of vitamin A deficiency. Medical Journal of Australia 188, 308; 108; Battacharjee, H., Das, K., Borah, R. R., Guha, K., Gogate, P., Purukayastha, S., Gilbert, C. (2008) Causes of childhood blindness in the northeastern states of India. Indian Journal of Ophthalmology 56, 495–499; Dole, K., Gilbert, C., Deshpande, M., Khandekar, R. (2009) Prevalence and determinants of xerophthalmia in preschool children in urban slums, Pune, India – a preliminary assessment. Ophthalmic Epidemiology 16, 8–14; Arlappa, N., Balakrishna, N., Laxmaiah, W., Raghu, P., Rao, V. V., Nair, K. M., Brahmam, G.N. (2011) Prevalence of vitamin A deficiency and its determinants among the rural pre-school children of Madhya Pradesh, India. Annals of Human Biology 38, 131–136; Arlappa, N, Laxmaiah, A., Balakrishna, N., Harikumar, R., Kodavanti, M. R., Gal Reddy, Ch., Saradkumar, S., Ravindranath, M., Brahmam, G. N. (2011) Micronutrient deficiency disorders among the rural children of West Bengal, India. Annals of Human Biology 38; 281–289; Arlappa, N., Venkaiah, K., Brahmam, G. N. (2011) Severe drought and the vitamin A status of rural pre-school children in India. Disasters 35, 577–586; Sinha, A., Jonas, J. B., Kulkarni, M., Nangia, V. (2011) Vitamin A deficiency in schoolchildren in urban central India: the Central India Children Eye Study. Archives of Ophthalmology 129, 1095–1096.

67 Gopalan, C. (2008) Vitamin A deficiency – overkill. Bulletin of the Nutrition Foundation of India 29 (3), 1–3.

68 West, K. P. Jr., Darnton-Hill, I. (2008) Vitamin A deficiency. In Semba & Bloem (2008), 377–433.

69 Semba (2007), 58–59; West (2008), 402–403.

70 International Institute for Population Sciences (IIPS) and Macro International. (2007) National Family Health Survey (NFHS-3), 2005–6 India, Vol. 1. Mumbai, IIPS, 41.

71 World Bank (1997) India: achievements and challenges in reducing poverty. Washington, D.C., World Bank.

72 Rajuladevi, A. K. (2001) Food poverty and consumption among landless labour households. Economic and Political Weekly 36, 2656–2664.

73 Dogra, B. (2007) Landless farmers demand a piece of the action. Inter Press Service News Agency, October 30, 2007; Anon (2007) India's poorest march on capital. Daily Times (Pakistan), October 28, 2007.

74 Gopalan, C. (1992) Vitamin A deficiency and child mortality. Lancet 340, 177–178.

75 Kapil, U. (2009) Time to stop giving indiscriminate massive doses of synthetic vitamin A to Indian children. Public Health Nutrition 12, 285–286; Mudur, G. S. (2008) Sheen goes off vitamin A effect on child deaths. The Telegraph, April 19, 2008.

76 Semba, R. D., Delange, F. (2008) Iodine deficiency disorders. In Semba & Bloem (2008), 507–529.

77 Fernandez, R. L. (1990) A simple matter of salt: an ethnography of nutritional deficiency in Spain. Berkeley, University of California Press.

78 The film was a disturbing portrait of peasants living in perpetual hunger in the rugged Sierra de Gata and featured close-ups of cretins, dwarves, and women with large goiters deforming their necks. The Franco regime censored the film, but Buñuel hoped that Marañon, in his prestigious position, could help procure authorization for distribution. During a private screening, however, Marañon was repulsed, and he reprimanded Buñuel for showing 'ugly things.' 'Why don't you show something nice,' the doctor suggested, 'like folk dances?' From Buñuel, L. (1984) My last sigh. New York, Random House, 141.

79 Marañon, G. (1928) Résumé de l'état actuel du problème du goitre endémique en Espagne. In Comptes-rendus de la Conférence Internationale du Goitre.Berne, 24–26 août 1927. Berne, Commission Suisse du Goitre and Hans Huber, 389–402.

80 Hunziker, H. (1924) Die Prophylaxe der grossen Schilddrüse. Bern and Leipzig, Ernst Bircher.

81 Escobar del Rey, F. (1985) Goitre and iodine deficiency in Spain. Lancet ii, 149–150.

82 Subcommittee for the Study of Endemic Goitre and Iodine Deficiency of the European Thyroid Association (1985) Goitre and iodine deficiency in Europe. Lancet i, 1289–1292.

83 Semba, R. D., de Pee, S., Sun, K., Campbell, A. A., Bloem, M. W., Raju, V. K. (2010) Low intake of vitamin A-rich foods among children, aged 12–35 months, in India: association with malnutrition, anemia, and missed child survival interventions. Nutrition 28, 956–962.

84 de Pee, S., West, C. E., Muhilal, Karyadi, D., Hautvast, J. G. A. J. (1995) Lack of improvement in vitamin A status with increased consumption of dark-green leafy vegetables. Lancet 346, 75–81.

85 Food and Nutrition Board, Institute of Medicine (2001). According to these guidelines, the vitamin A bioavailability is 12:1 for β-carotene from vegetables and fruits (12 μg β-carotene = 1 retinol activity equivalent (RAE) and 24:1 for other provitamin A carotenoids (24 μg α-carotene or β-cryptoxanthin = 1 RAE).

86 Haskell, M. J., Jamil, K. M., Hassan, F., Peerson, J. M., Hoosain, M. I., Fuchs, G. J., Brown, K. H. (2004) Daily consumption of Indian spinach (*Basella alba*) or sweet potatoes has a positive effect on total-body vitamin A stores in Bangladeshi men. American Journal of Clinical Nutrition 80, 705–714; Tang, G., Qin, J., Dolnikowsi, G. G., Russell, R. M., Grusak, M. A. (2005) Spinach or carrots can supply significant amounts of vitamin A as assessed by feeding with intrinsically deuterated vegetables. American Journal of Clinical Nutrition 82, 821–828; Khan, N. C., West, C. E., de Pee, S., Bosch, D., Phuong, H. D., Hulshof, P. J. M., Khoi, H. H., Verhoef, H., Hautvast, J. G. A. J. (2007) The contribution of plant foods to the vitamin A supply of lactating women in Vietnam: a randomized controlled trial. American Journal of Clinical Nutrition 85, 1112–1120.

87 The Recommended Dietary Allowance (RDA) of vitamin A for children, aged one to three years is 300 µg RAE/day. Using the conversion factors of the Institute of Medicine (12:1 for β-carotene and 24:1 for α-carotene and β-cryptoxanthin), the child would need to eat 64 grams of dark green leafy vegetables per day, or about four servings, given a portion size of 15 grams. However, the conversion factor is probably closer to 24:1 for dark green leafy vegetables [Tang, 2005; Khan, 2007], in which case a child who need to consume eight 15 gram servings of dark green leafy vegetables per day.

88 West (2008), 385.

Night Blindness Among Black Troops and White Troops in the US Civil War

Night blindness was common during the US Civil War, as reported in chapter 4. The United States Surgeon General's Office published the monthly reports of morbidity and mortality in volume 1 of *The Medical and Surgical History of the War of the Rebellion (1861–1865)*. The data used for these analyses were the number of cases per month of night blindness, acute diarrhea, and chronic diarrhea, and deaths due to acute diarrhea and chronic diarrhea, relative to total troop strength per month. Analyses were restricted to the Atlantic and Central regions for both black and white troops. The Atlantic region consisted of seven armies or departments: East, Middle, Washington, Army of the Potomac, Virginia, North Carolina, and South. The Central region consisted of nine departments: West Virginia, Northern, Ohio, Cumberland, Tennessee, Gulf, Northwest, Missouri, and Arkansas. Univariate linear regression models were used to examine the relationship between monthly point prevalence of night blindness and monthly point prevalence of acute and chronic diarrhea and monthly case fatality rates for acute and chronic diarrhea. The monthly point prevalence is the proportion of soldiers out of the total number of soldiers with the particular condition during each month. The case fatality rate is the proportion of soldiers affected with the condition that died, calculated for each month. Analyses were conducted using SAS version 9.1 (SAS Institute, Cary, N.C., USA).

The period July 1863 to June 1865 was chosen for the analysis. Data were reported for both black and white troops during this period. The monthly point prevalence of night blindness among troops was higher in black troops than white troops, and both showed a seasonal pattern with a peak prevalence in the summer (chapter 4; textbox 4–1). The point prevalence of acute and chronic diarrhea also showed a seasonal pattern with peaks in the summer. The case fatality rates for both acute and chronic diarrhea were higher among black troops than white troops, and the case fatality rates were much higher for chronic diarrhea than acute diarrhea. The peak in case fatality for acute diarrhea occurred in the summer, while that for chronic diarrhea peaked in the autumn and winter.

For black troops, the monthly point prevalence of night blindness was associated with the monthly point prevalence of diarrhea (p = 0.005) and chronic diarrhea (p =

0.015). For white troops, the monthly point prevalence of night blindness was associated with the monthly point prevalence of diarrhea (p < 0.0001) and chronic diarrhea (p = 0.005). In all troops together, the monthly point prevalence of night blindness was associated with the monthly case fatality rate for diarrhea (p = 0.002), but not chronic diarrhea (p = 0.18).

These results are suggestive and should be interpreted with caution, since the individual medical records of soldiers were not linked with other data such as age, demographic information, weight, height, and other detailed data. It is not known whether soldiers who developed night blindness were also the same individuals who had diarrhea, but it is well established from epidemiological studies that diarrhea and night blindness have a strong association. However, these data clearly show that the rates of night blindness, acute diarrhea, and case fatality rates from acute and chronic diarrhea were higher among black troops than white troops during the US Civil War.

Bibliography

Manuscript Sources

I.   Caird Library, National Maritime Museum, Greenwich, United Kingdom
     JOD/10   Journals and Diaries 1781. Books of menus, kept by John
                  Gulivar, Adm Digby's steward
II. Connecticut Agricultural Experiment Station, New Haven, Connecticut,
     USA
     TBO   Papers of Thomas Burr Osborne
III. Katz, Joanne, Johns Hopkins University, Baltimore, Maryland, USA
     JK   Personal files and correspondence from Albay Mother and
                  Child Health Project
IV. National Archives, Kew, United Kingdom
     FD 1/3790   Medical Research Committee and Medical Research Council.
                  Nutritive value of milk (Corry Mann)
     FD 1/3791   Medical Research Committee and Medical Research Council.
                  Nutritive value of milk II
     FD 1/3792   Medical Research Committee and Medical Research Council.
                  Nutritive value of milk III
V.   Kenneth Spencer Research Library, University of Kansas Libraries, Lawrence,
     Kansas, USA
     PP120   Elmer V. McCollum Collection, University Archives
VI. Wellcome Institute for the History of Medicine, Contemporary Medical
     Archives Centre, London, United Kingdom
     PP/MEL   Papers of Sir Edward Mellanby (1884–1955) and Lady (May)
                  Mellanby (1882–1978).
     PP/CDW   Papers of Cicely Delphine Williams (1893–1992)
VII. University of Wisconsin – Madison Archives, Steenbock Library, Madison,
     Wisconsin, USA
     9/1/1/22-1   Harry L. Russell. Diaries. Black Books.
     9/11/13-2   Harry Steenbock papers, 1905–1960

VIII.   Yale University Library, Manuscripts and Archives, New Haven, Connecticut,
        USA
        MS1146   Lafayette Benedict Mendel Papers.

Published Sources

The following list of primary and secondary literature is meant to highlight some of
the main works related to vitamin A and other topics treated in this book. The bibli-
ography cited here is not mean to be comprehensive or exhaustive, as greater detail is
found within the references for each chapter.

Arroyave, G., Aguilar, J. R., Flores, M., Guzmán, M. A. (1979) *Evaluation of sugar fortification with
   vitamin A at the national level.* Scientific Publication 384. Washington, D. C., Pan American
   Health Organization and World Health Organization.
Beaton, G. H., Martorell, R., L'Abbé, K. A., Edmonston, B., McCabe, G., Ross, A. C., Harvey, B.
   (1992) *Effectiveness of vitamin A supplementation in the control of young child morbidity and
   mortality in developing countries.* Toronto, University of Toronto.
Benoiston de Chateauneuf, [L. F.]. (1830) De la durée de la vie chez le riche et chez le pauvre.
   Annales d'hygiène publique et de médecine légale 3, 5–15.
Billard, C. (1828) *Traité des maladies des enfans nouveax-nés et à la mamelle, fondé sur de nouvelles
   observations cliniques et d'anatomie pathologique, faites a l'Hôpital des Enfans-Trouvé de Paris,
   dans le service de M. Baron.* Paris, J. B. Baillière.
Bitot, [P]. (1863) Mémoire sur une lésion conjonctivale non encore décrite, coïncidant avec
   l'héméralopie. *Gazette hebdomadaire de médecine et de chirurgie* 10, 284–288.
Blegvad, O. (1924) Xerophthalmia, keratomalacia and xerosis conjunctivae. *American Journal of
   Ophthalmology* 7, 89–117.
Bloch, C. E. (1921) Clinical investigation of xerophthalmia and dystrophy in infants and young
   children (*xerophthalmia et dystrophia alipogenetica*). *Journal of Hygiene* 19, 283–304.
Bollett, A. J. (2002) *Civil War medicine: challenges and triumphs.* Tucson, Galen Press.
Carpenter, K. (1994) *Protein and energy: a study of changing ideas in nutrition.* Cambridge,
   Cambridge University Press.
Carr, F. H., Price, E. A. (1926) Colour reactions attributed to vitamin A. *Biochemical Journal* 20,
   497–501.
Chaussonnet, M. L. E. (1870). *De l'héméralopie aiguë.* Thèse pour le doctorat en médecine.
   Faculté de médecine de Paris, no. 63. Paris, A. Parent.
Chevalier, L. (1973) *Laboring classes and dangerous classes in Paris during the first half of the
   nineteenth century.* New York, Howard Fertig.
Corry Mann, H. C. (1926) Diets for boys during the school age. *Medical Research Council Special
   Report Series No. 105.* London, His Majesty's Stationery Office.
Decaisne, E. (1871) Des modifications que subit le lait de femme par suite d'une alimentation
   insuffisante. Observations recueillies pendant le siége de Paris. *Comptes-rendus
   hebdomadaires des séances de l'Académie des Sciences* 73, 128–131.
de Haas, J. H. (1931) On keratomalacia in Java and Sumatra (in particular upon the Karo-
   Plateau) and in Holland. *Mededeelingen dienst der volksgezondheid in Nederlandsch-Indië* 20,
   1–11.

de Haas, J. H., Posthuma, J. H., Meulemans, O. (1940) Xerophthalmie bij kinderen in Batavia. *Geneeskundig tijdschrift voor Nederlandsch-Indië* 80, 928–950.

de Pee, S., West, C. E., Muhilal, Karyadi, D., Hautvast, J. G. A. J. (1995) Lack of improvement in vitamin A status with increased consumption of dark-green leafy vegetables. *Lancet* 346, 75–81.

Drummond, J. C. (1920) The nomenclature of the so-called accessory food factors (vitamins). *Biochemical Journal* 14, 660.

Dumas [J. B. A.] (1871) Note sur la constitution du lait et du sang. *Le moniteur scientifique* 3 ser., 1, 778–783.

Dupoux, A. (1958) *Sur les pas de Monsieur Vincent. Trois cents ans d'histoire parisienne de l'enfance abandonée.* Paris: Revue de l'Assistance Publique à Paris.

Ellison, J. B. (1932) Intensive vitamin therapy in measles. *British Medical Journal* 2, 708–711.

Fuchs, R. G. (1984) *Abandoned children: foundlings and child welfare in nineteenth-century France.* Albany, State University of New York Press.

Gayet. (1888) Héméralopie. In Dechambre, A. *Dictionnaire encyclopédique des sciences médicales.* 4 series, vol. 13. Paris, G. Masson, pp. 145–177.

Goldberger, J. (1916) Pellagra: causation and a method of prevention. A summary of some of the recent studies of the United States Public Health Service. *Journal of the American Medical Association* 60, 471–476.

Green, H. N., Mellanby, E. (1928). Vitamin A as an anti-infective agent. *British Medical Journal* 2, 691–696.

Green, H. N., Pindar, D., Davis, G., Mellanby, E. (1931) Diet as a prophylactic agent against puerperal sepsis. *British Medical Journal* 2, 595–598.

Holmes, H. N., Corbet, R. E. (1937) The isolation of crystalline vitamin A. *Journal of the American Chemical Society* 59, 2042–2047.

Hopkins, F. G. (1912) Feeding experiments illustrating the importance of accessory factors in normal dietaries. *Journal of Physiology* 44, 425–460.

Hume, E. M., Krebs, H.A. (1949) *Vitamin A requirements of human adults: an experimental study of vitamin A deprivation in man. A report of the Vitamin A Sub-Committee of the Accessory Food Factors Committee.* Privy Council, Medical Research Council Special Report Series No. 264. London, His Majesty's Stationery Office.

Hussey, G. D., Klein, M. (1990) A randomized, controlled trial of vitamin A in children with severe measles. *New England Journal of Medicine* 323, 160–164.

Isler, O., Huber, W., Ronco, A., Kofler, M. (1947) Synthese des Vitamin A. *Helvetica Chimica Acta* 30, 1911–1921.

Karrer, P., Helfenstein, A., Wehrli, H., Wettstein, A. (1930) Über die Konstitution des Lycopins und Carotins. *Helvetica Chimica Acta* 13, 1084–1099.

Karrer, P., Morf, R. (1931) Zur Konstitution des β-Carotins und β-Dihydro-carotins. *Helvetica Chimica Acta* 14, 1033–1036.

Karrer, P., Morf, R., Schöpp, K. (1931) Zur Kenntnis des Vitamins-A aus Fischtranen. *Helvetica Chimica Acta* 14, 1036–1040, 1431–1436.

Kranzberg, M. (1950) *The siege of Paris, 1870–1871: a political and social history.* Ithaca, Cornell University Press.

Lomborg, B. (ed.) (2004) *Global crises, global solutions.* Cambridge, Cambridge University Press.

Lunin, N. (1881) Über die Bedeutung der anorganischen Salze für die Ernährung des Thieres. *Zeitschrift für physiologische Chemie* 5, 31–39.

Magendie, F. (1816) Mémoire sur les propriétés nutritives des substances qui ne contiennent pas d'azote. *Bulletin des sciences par la Société Philomatique de Paris* 4, 137–138.

Magendie, F. (1816) Mémoire sur les propriétés nutritives des substances qui ne contiennent pas d'azote. *Annales de chimie et de physique* (sér. 2) 3, 66–77.

Magendie, F. (1841) Rapport fait à l'Académie des Sciences au nom de la Commission dite de la gélatine. *Compte-rendus des séances de l'Académie des Sciences* 13, 237–283.

McCollum, E. V., Davis, M. (1913) The necessity of certain lipins in the diet during growth. *Journal of Biological Chemistry* 15, 167–175.

McCollum, C. V., Simmonds, N., Becker, J. E., Shipley, P. G. (1922) Studies on experimental rickets. XXI. An experimental demonstration of the existence of a vitamin which promotes calcium deposition. *Journal of Biological Chemistry* 53, 293–312.

Mellanby, H. N., Green, H. N. (1929) Vitamin A as an anti-infective agent. Its use in the treatment of puerperal septicaemia. *British Medical Journal* 1, 984–986.

Mendes, J. C. (1862) Estudo sobre a hemeralopia a propósito dos casos observados na guarnição de Lisboa. *Escholiaste medico* 13, 22–24, 39–42, 55–58, 70–72, 85–88, 106–110.

Moore, T. (1929) The relation of carotin to vitamin A. *Lancet* 2, 380–381.

Mori, M. (1904) Über den sog. Hikan (Xerosis conjunctivae infantum ev. Keratomalacie). II. Mitteilung. *Jahrbuch für Kinderheilkunde und physische Erziehung* 59, 175–195.

Netter, A. (1862–1863) Nouveau mémoire sur l'héméralopie épidémique et le traitement de cette maladie par les cabinets ténébreux. *Gazette Médicale de Strasbourg* 22, 164–171, 186–192; 23, 9–17, 21–27.

Oomen, H. A. P. C. (1953) Infant malnutrition in Indonesia. *Bulletin of the World Health Organization* 9, 371–384.

Osborne, T. B., Mendel, L. B. (1911) *Feeding experiments with isolated food-substances.* Washington, D.C., Carnegie Institute of Washington, Publication No. 156.

Osborne, T. B., Mendel, L. B. (1913) The relationship of growth to the chemical constituents of the diet. *Journal of Biological Chemistry* 15, 311–326.

Ramalingaswami, V. (1948) Nutritional diarrhoea due to vitamin A deficiency. *Indian Journal of Medical Sciences* 2, 665–674.

Scrimshaw, N. S., Taylor, C. E., Gordon, J. E. (1968) *Interactions of nutrition and infection.* Geneva, World Health Organization.

Socin, C. A. (1891) In welcher Form wird das Eisen resorbirt? *Zeitschrift für physiologische Chemie* 15, 93–139.

Sommer, A., Hussaini, G., Tarwotjo, I., Susanto, D. (1983) Increased mortality in children with mild vitamin A deficiency. *Lancet* 2, 585–588.

Sommer, A., Tarwotjo, I., Djunaedi, E., West, K. P., Jr., Loeden, A. A., Tilden, R., Mele, L., and the Aceh Study Group. (1986) Impact of vitamin A supplementation on childhood mortality. A randomized controlled community trial. *Lancet* 1, 1169–1173.

Steenbock, H. (1924) The induction of growth promoting and calcifying properties in a ration by exposure to light. *Science* 60, 224–225.

Stepp, W. (1911) Experimentelle Untersuchungen über die Bedeutung der Lipoide für die Ernährung. *Zeitschrift für Biologie* 57, 135–170.

United States. Surgeon-General's Office. (1870) *The medical and surgical history of the war of the rebellion (1861–65). Prepared under the direction of the Surgeon General Joseph K. Barnes, United States Army.* Vol. 1. Part 1. Washington, D.C., Government Printing Office.

Villermé, L. R. (1830) De la mortalité dans les divers quartiers de la ville de Paris, et des causes qui la rendent très différente dans plusieurs d'entre eux, ainsi que dans les divers quartiers de beaucoup de grandes villes. *Annales d'hygiène publique et de médecine légale* 3, 294–341

von Euler, B., von Euler, H., Hellström, H. (1928) A-Vitaminwirkungen der Lipochrome. *Biochemische Zeitschrift* 203, 370–384.

Wald, G. (1933) Vitamin A in the retina. *Nature* 132, 316–317.

Wolbach, S. B., Howe, P. R. (1925) Tissue changes following deprivation of fat soluble A vitamin. *Journal of Experimental Medicine* 42, 753–777.

World Bank. (1993) *World development report 1993: investing in health.* New York, World Bank and Oxford University Press.

Subject Index

United States Agency for International
    Development 161

Vaccines
    diphtheria-pertussis-tetanus 174
    oral poliovirus 17
Val de Grâce Hospital, Paris 31
Vegetable vitamin A 189, 190
Vesicatories 3, 57
Vietnam 184
Villermé, Louis-René 26–28
Visual cycle 99
Vitamin A
    anti-infective therapy 132–148
    antimony trichloride assay 93, 94
    bioavailability in vegetables and fruit 189,
        190
    capsules 161, 170, 173
    carotene relationship 93, 94, 98, 99
    crystallization 99
    risk-benefit 171, 174
    structure 98, 99
    sugar fortification 160, 181, 182
    treatment of measles, *see* Measles
    yellow color in foods 93
Vitamin C 75, 84
Vitamin D 68, 74, 75, 94
Vitamines 77
Vitamin K 159
Voit, Carl von 67

Wald, George 94, 99
Water-soluble B 73, 74, 77
West, Clive 189

Western Hemisphere Nutrition Congress 159
West, Keith P., Jr. 177, 179
Whooping cough 140, 146
Widmark, Erik 128
Wilhelm II, King 46
Willcock, Edith 70
Williams, Cicely 155
Williamson, Alexander Dewar 154
Williams, Robert 74
Willie, W.A. 151
Wilmer Eye Institute 161
Windaus, Adolf 92, 96, 97
Wisconsin Agricultural Experiment Station 87
Wisconsin Alumni Research Foundation 98
Wolbach, S. Burt 128
Wolcott, Roger 52
Women's Foundation for Health 147
World Bank 177, 184
World Declaration for the Survival, Protection,
    and Development of Children 181
World Food Conference 161
World Health Organization 42, 155–157, 181
World Summit for Children 181
World War I 110, 143
World War II 139, 156

Xerophthalmia 45, 77, 83, 94, 108–110, 112,
    152, 157, 158, 162, 184

Year without a summer 26, 26
Yellow fever 8